Bat Surveys

Good Practice Guidelines

Editorial Board

Chair: Katie Parsons

Richard Crompton

Richard Graves

Steve Markham

Jean Matthews

Mike Oxford

Peter Shepherd

Sandie Sowler

Bat Conservation Trust

15 Cloisters House, 8 Battersea Park Road

London SW8 4BG, UK

Registered Charity No: 1012361 Company Limited by Guarantee.
Registered in England No: 2712823. Vat Reg No: 877158773.

www.bats.org.uk

Contents

Foreword

The practical implementation of bat conservation is evolving rapidly. It is clear that, where development or land management proposals have the potential to impact on bats, there needs to be quality control in the standard of bat survey undertaken and that the survey should be appropriate to the situation.

Bat Surveys - Good Practice Guidelines provides a guide to the type and level of survey required for different habitats, in order to adequately ascertain their use by bats. These guidelines were developed with input from experts in the field. The Bat Conservation Trust (BCT) organised workshops and consultations to gather new knowledge and gain consensus about survey methods and the effort required to have confidence in a survey's results. These guidelines provide the best current thinking on bat surveys and future editions will be produced as knowledge about bats improves and survey methods develop.

My thanks go to all who were involved with the production of this document. I hope that by bringing together their expertise, bats and the habitats they require will be better conserved now and in the future.

Dr Katie Parsons, Head of Biodiversity, BCT

Message from BCT's Chief Executive Officer

This document would not have been possible without the goodwill and enthusiasm of bat experts throughout the UK and Europe. I have enjoyed working with everyone to produce these guidelines and am confident that they form a first step towards consistent, high quality bat surveys to benefit bat conservation.

Amy Coyte, BCT

Acknowledgements

This publication has benefited greatly from the input of a large number of people. In particular, the Bat Conservation Trust would like to acknowledge the major contribution by Katie Parsons as an author and as Chair of the Editorial Board. We would like to say a big thank you to other members of the Editorial Board, who were Richard Crompton, Richard Graves, Steve Markham, Jean Matthews, Mike Oxford, Peter Shepherd and Sandie Sowler. Each gave significant amounts of their time, free of charge, to pull the document together and drive it forward to completion.

We are grateful to Gail Armstrong, Dennis Blackwell, Conor Kelleher, Simon Mickleburgh, Kit Stoner and Linda Yost for their editorial input and for proofreading.

We would like to thank the participants of two workshops on bat surveys that formed the basis of these guidelines – Jules Agate, Jim Alexander, Lothar Bach, Andrew Baker, Geoff Billington, Peter Boye, Patty Briggs, Robert Brinkman, Colin Catto, Amy Coyte, Richard Crompton, Ian Davidson-Watts, Johnny de Jong, Marcus Dietz, Matthew Dodds, John Drewett, Frank Greenaway, Tony Hutson, Gareth Jones, Conor Kelleher, Herman Limpens, Iain Mackie, Tony Martin, Jean Matthews, Brenda Mayle, Tony Mitchell-Jones, Martin Noble, Alison Rasey, Paola Reason, Phil Richardson, Mark Robinson, Tony Sangwine, Gillie Sargent, Peter Shepherd, Peter Smith, Kit Stoner, Stuart Wilson and Stephanie Wray. BCT is grateful to Nigel Westerway Associates who facilitated the three-day workshop in December 2004 and to Baker Shepherd Gillespie and Cresswell Associates Ltd. who produced workshop reports. Many of these people also helped the sub-editors with the wording and content of specific chapters, as did Andrew Cowan, John Drewett, Michael Ford, Alison Fure, Doug Ireland, Andy Kendall, Mark Robinson, Jon Russ, David Webb and SLR Consulting Ltd.

BCT would also like to thank all those who responded to the consultation held on the draft document between February and March 2007. All comments and suggestions were considered by the Editorial Board and many were incorporated into this version. Space precludes us from listing everyone here but their input was invaluable.

Several members of BCT staff have contributed a great deal to this document in a variety of ways: Michael Fray in handling the consultation responses as they came in; and Jules Agate, Jaime Eastham, Karen Haysom, Simon Mickleburgh, Alison Rasey, Kit Stoner and Carol Williams in providing a sounding board and taking time to debate sections of the text and the use of terminology.

We would like to thank the following organisations for supporting the production of this document by funding workshops and the Head of Biodiversity's time on this project: Environment Agency; Esmée Fairbairn Foundation; Forestry Commission; Highways Agency; and Natural England. Many thanks to Faber Maunsell for providing funding towards the printing costs, thereby enabling proceeds from the sale of this document to go towards bat conservation.

Many others have helped with the production of this document and it has not been possible to list everyone by name. We would like to thank you all for your time and expertise.

Chapter 1

Introduction

1.1 Aim of this document

To conserve bats effectively, and to meet national and international statutory obligations and agreements, there is a clear need for quality controls and appropriate levels of bat survey to be established. To date, there has been uncertainty around the type of survey and level of survey effort needed, when to survey, and how these decisions are affected by the habitat, time of year, geographical location and the species present or likely to be present in the prospective survey area.

This document provides good practice guidance for those commissioning, undertaking or reviewing bat surveys throughout the UK. It is intended to bring about improvements in the standard and consistency of bat surveys and survey reports and ultimately lead to greater understanding about bats and improvements in their protection and conservation.

The guidance will need to be interpreted and adapted on a case-by-case basis, according to the expert judgement of those involved. There is no substitute for knowledge and experience in survey planning, methodology and interpretation of findings, and these guidelines intend to support these. Where examples are given they are descriptive rather than prescriptive.

1.2 Who should use this document

The guidelines are for:

- ecologists (such as professional consultants) undertaking bat surveys;
- developers commissioning bat surveys from professional consultants in order to provide information in support of a valid planning application;
- planners and ecologists working in local authorities, Statutory Nature Conservation Organisations (SNCOs) and non-governmental organisations who are responsible for reviewing and assessing the implications of bat survey results;
- nature conservationists with an interest in, and responsibility for, the conservation of bats – for example, bat workers undertaking voluntary roost visits; and
- ecologists conducting ecological research into the behaviour, occurrence and distribution of bats.

1.3 Layout

The document starts by assessing the purpose of, and need for, a bat survey. This is followed by preparation and planning for surveys, and then by a chapter on each type of survey. Additional information can be found in the Appendices.

This guidance will be most useful if read in its entirety and then used as a reference to be 'dipped' into as required.

The document is available to purchase in hard copy; contact the Bat Conservation Trust for details. It is also available electronically to be downloaded from: **www.bats.org.uk/biodiversity/ BatSurveyGuidelines.asp**

We intend to review and update this document as new information becomes available, so please check the website regularly to ensure that you have the latest version.

Case studies to illustrate the process of conducting bat survey in different scenarios will be made available on the website and periodically reviewed and updated.

1.4 Background

The first draft of this document was prepared following input from UK and European bat experts via a series of workshops. Each chapter was then sub-edited by a member of the Editorial Board and a draft put out to public consultation. Responses were received from a wide variety of end-users including ecological consultants, volunteer bat workers, local authority ecologists, developers, SNCOs and government departments. Following this consultation process, the Editorial Board produced the final document presented here.

1.5 Context

1.5.1 Legislative context

Annex II of the *Council Directive 92/43/EEC 1992 on the Conservation of Natural Habitats and of Wild Fauna and Flora* (EC Habitats Directive) lists animal and plant species of Community interest, the conservation of which requires the designation of Special Areas of Conservation (SACs); Annex IV lists animal and plant species of Community interest in need of strict protection. All bats are listed in Annex IV; some bats are listed in Annex II.

The breeding sites and resting places of bats are usually known as 'roosts', of which there are a variety of types (see Chapter 6). Resting places also include, for example, feeding perches where a bat rests while waiting to catch or while consuming, prey. Bat roosts are protected even when bats are not present.

No offence is committed if work is done under, and in accordance with, a licence issued under the Habitats Regulations. A licence may be issued for scientific or educational purposes, conserving wild animals, preventing the spread of disease or another imperative reason of overriding public interest. The licensing authority must be satisfied that there is no satisfactory alternative to the proposed action and that it will not be detrimental to the Favourable Conservation Status (FCS) of the bats.

Box 1.1 Summary of the main pieces of UK legislation pertaining to the protection of bats

	Habitats Regulations (transposing the EC Habitats Directive)	**Other nature conservation legislation**
England and Wales	Conservation (Natural Habitats, &c.) Regulations 1994 (Statutory Instrument (SI) 1994 No. 2716) as amended by SI 1997 No. 3055 and SI 2000 No. 192.	Wildlife and Countryside Act 1981 as amended Countryside and Rights of Way Act 2000 Natural Environment and Rural Communities Act 2006
Northern Ireland	The Conservation (Natural Habitats, etc.) Regulations (Northern Ireland) 1995 (Statutory Rule (SR) 1995 No. 380) as amended by SR 2004 No. 435.	Wildlife (Northern Ireland) Order 1985 as amended Nature Conservation and Amenity Lands (Northern Ireland) Order 1985 as amended Environment Order 2002
Scotland	Conservation (Natural Habitat &c) Regulations 1994 as amended by SI 1997 No. 3055, SI 2000 No. 192, Scottish Statutory Instrument (SSI) 2004 No. 475 and SSI 2007 No. 80.	Wildlife and Countryside Act 1981 as amended The Nature Conservation (Scotland) Act 2004

In the UK[1], the EC Habitats Directive has been transposed into national laws by means of the *Conservation (Natural Habitats, &c.) Regulations 1994* (as amended) and the *Conservation (Natural Habitats,* etc.*) Regulations (Northern Ireland) 1995* (as amended). These are commonly and collectively known as the 'Habitats Regulations'. For the purposes of this document, where referred to in the text, they will be called the 'Habitats Regulations'. A summary of the relevant nature conservation legislation is shown in Box 1.1. Always consult the appropriate Habitats Regulations before surveying/ operating in each part of the UK and in other EC member countries.

The Habitats Regulations transpose the EC Habitats Directive to give bats, their breeding sites and resting places a high level of strict protection. In summary, it is a criminal offence (subject to certain specific exceptions outlined below) to:

- capture or kill a bat;
- disturb a bat whilst in a place of shelter or rest; or
- damage or destroy a bat's breeding site or resting place.

Two main types of licences are of relevance to this document:

1. Licences needed by trained bat surveyors or researchers to allow them to undertake certain activities or projects that would otherwise contravene the Habitats Regulations; for example, entry into a bat roost or capture and radio-tracking of bats. These licences are dealt with in more detail in Section 3.4.

2. Habitats Regulations licences (sometimes called European Protected Species (EPS) licences) permit work that would otherwise constitute an offence by disturbing bats and/or damaging or destroying their roosts; this might include demolishing a building containing a bat roost, converting a barn or felling a tree. In each case the roost is either destroyed or damaged. A Habitats Regulations licence application consists of an application form, a Method Statement and a Reasoned Statement.

1 The Habitats Directive does not apply to the Isle of Man and the Channel Islands, which are part of the British Isles but not of the United Kingdom. The relevant legislation in the Isle of Man and the Channel Islands should be consulted.

The Method Statement is prepared by the ecologist and comprises survey information, assessment and mitigation/enhancement measures. These measures are designed to establish that *"the action authorised will not be detrimental to the maintenance of the population of the species concerned at a favourable conservation status in their natural range"* (Habitats Regulations 44(3)(6)).

1.5.2 Planning policy context

Government planning policy guidance throughout the UK[2] requires local planning authorities to take account of the conservation of protected species when determining planning applications. This makes the presence of a protected species a material consideration when assessing a development proposal that, if carried out, would be likely to result in harm to the species or its habitat.

In the case of European Protected Species such as bats, planning policy emphasises that strict statutory provisions apply, to which a planning authority must have due regard (e.g. Habitats Regulations 3(4) and see also Section 2.2.3).

For instance, *"It is essential that the presence or otherwise of protected species, and the extent that they may be affected by the proposed development, is established before the planning permission is granted, otherwise all relevant material considerations may not have been addressed in making the decision* (ODPM Circular 06/2005; paragraph 99)".[3]

This requirement has important implications for bat surveys as it means that, where there is a reasonable likelihood of the species being present and affected by the development, surveys should be carried out before planning permission is considered.

The British Standards Institution has published a Publicly Available Specification (PAS), *PAS 2010 Planning to Halt the Loss of Biodiversity: Biodiversity Conservation Standards for Planning in the UK - Code of Practice* (British Standards Institution, 2006) as a code of practice for planning authorities on how they should address biodiversity in their planning work. PAS 2010 identifies where planning authorities have clear responsibilities for biodiversity conservation. It specifies tasks that should be undertaken to discharge their planning functions in a manner that is compliant with statutory obligations, government policy and good practice. It provides a consistent framework by which planning authorities can take effective action to ensure, through forward planning, development control and management decisions, that they are able to help halt the loss of biodiversity and thus contribute to sustainable development.

1.5.3 Biodiversity policy context

Bats were given legal protection in the UK because of evidence that bat populations have declined significantly in the last century. The Convention on Biological Diversity, signed in Rio de Janeiro, Brazil in 1992, requires Parties to develop national strategies and to undertake a range of actions aimed at maintaining or restoring biodiversity. The UK Biodiversity Strategy was produced in response to the Convention.

Individual Species Action Plans (SAPs) have been developed to address the causes of decline for those species that have been identified as priorities for UK conservation action. Country-level lists contain species considered of national importance in biodiversity strategies[4]. Local Biodiversity Action Plans (LBAPs) have been produced for additional species of local importance. Details of SAPs and LBAPs are available from **www.ukbap.org.uk**.

In England and Wales, the Natural Environment and Rural Communities (NERC) Act, 2006 imposes a duty on all public bodies, including local authorities and statutory bodies, in exercising their functions, *"to have due regard, so far as is consistent with the proper exercise of those functions, to the purpose of conserving biodiversity"* [Section 40 (1)]. It notes that *"conserving biodiversity includes restoring or enhancing a population or habitat"* [Section 40 (3)].

In Scotland, the Nature Conservation (Scotland) Act 2004 imposes a duty on every public body and office holder, in exercising any function, *"to further the conservation of biodiversity so far as is consistent with the proper exercise of those functions"* [Section 1 (1)].

2 Government policy guidance for biodiversity and nature conservation throughout the UK is provided in the following planning documents:
In England: Planning Policy Statement (PPS) 9 and ODPM Circular 06/2005.
In Northern Ireland: Planning Policy Statement 2.
In Wales: Technical Advice Note (TAN) 5.
In Scotland: National Planning Policy Guidance (NPPG) 14 and Planning Advice Note (PAN) 60 PPS 2.

3 The principle of ensuring that the planning authority has sufficient information prior to determination has been established through judicial review: *'Cornwall Case'*: R v Cornwall County Council *ex parte* Jill Hardy 2001 Journal of Planning Law 786. Although this case strictly refers to statutory EIA as planning permission was overturned in relation to the EIA Regulations that require full environmental information to be presented, this has been interpreted as applicable to all cases including assessments not subject to the EIA Regulations.

4 The country-level biodiversity lists are found as follows:
In England: The *List of Habitats and Species Important to Biological Diversity in England* was prepared under Section 74(2) of the Countryside and Rights of Way Act 2000.
In Northern Ireland: The *List of Northern Ireland Priority Species* was prepared following a recommendation contained in the report *Biodiversity in Northern Ireland: Recommendations to Government for a Biodiversity Strategy* by the Northern Ireland Biodiversity Group.
In Scotland: The *List of Species and Habitats Considered to be of Principal Importance for the Purpose of Biodiversity Conservation in Scotland* was prepared under Section 2(4) of the Nature Conservation (Scotland) Act 2004.
In Wales: The *List of Species of Principal Importance for the Conservation of Biological Diversity* was prepared under Section 74 of the Countryside and Rights of Way Act 2000.

1.5.4 Other guidance

English Nature[5]'s *Bat Mitigation Guidelines* (Mitchell-Jones, 2004) focus on mitigation for bat roosts in buildings, but also address a wide variety of topics including legislation and licensing, and include a short section on survey objectives, methods and standards.

The third edition of the *Bat Workers' Manual* (Mitchell-Jones and McLeish, 2004) covers all aspects of bat work and is an essential reference for anyone wishing to obtain a bat licence in the UK. Of particular interest to readers of these guidelines are Chapter 3, which contains an overview of the types of methodologies that can be used for survey and monitoring for conservation purposes; and Chapter 4, which gives detail on techniques for catching bats.

For guidance on impact assessment see *Guidelines for Ecological Impact Assessment in the United Kingdom* (Institute of Ecology and Environmental Management, 2006) and *Guidelines for Baseline Ecological Assessment* (Institute of Environmental Assessment, 1995, currently under review).

Various guidance has been, and is being, produced by the Scientific Working Group to The Agreement on the Conservation of Populations of European Bats (EUROBATS). The UK Government is a Party to the Agreement, which was ratified in January 1994 following a resolution adopted by the first Conference of Parties of the Convention on Migratory Species (CMS) to develop an Agreement to protect all European bats. EUROBATS sets legal protection standards, while developing and promoting trans-boundary conservation and management strategies, research, and public awareness across the Agreement area (48 Range States). Further information about the Agreement, along with adopted Resolutions and guidance can be found on the EUROBATS website **www.eurobats.org**. Particular Resolutions or guidance from EUROBATS are referenced in the appropriate place within this document.

5 English Nature joined with the Countryside Agency and the Rural Development Service to form Natural England on 2 October 2006. Any publications published by English Nature before that date are attributed to English Nature, and afterwards to Natural England.

Chapter 2

Assessing the need for bat survey

2.1 What is a survey?

Surveying is a scientific activity during which people and/or equipment gather discrete information from a site or area. Survey information should provide facts: objectives, methods, data (lists, tables, graphs, plans etc.), along with any limitations.

In these guidelines, a survey is defined as a sampling activity that uses a wide range of variables to describe a site or an area, and is carried out as a time-bound once-only project, although the survey may involve repeated visits or observations. It is distinguishable from monitoring, which involves repeated sampling, either year-on-year or periodically, usually to assess whether a particular objective or standard has been attained (definition adapted from **www.jncc.gov.uk**).

Bat surveys are undertaken for a variety of reasons. The most basic type of survey will usually consist of the collation of existing data and a walkover survey to determine if there are any features or habitats that could potentially support bats. These basic surveys are often referred to as *scoping* or *first stage surveys* (see Chapter 3).

The objectives of a survey will influence the type of *second stage survey* that is undertaken, the methods used and the amount of survey effort needed (see Chapters 4-10).

Survey information should indicate:

- the species present or likely to be present at the site or in the area;
- the approximate numbers of individuals and/or the relative importance of the population in a local, regional and national context; and
- when, and for what purposes, bats use the site.

2.2 Reasons for surveying

Surveying for bats may be undertaken for a variety of reasons but most will fall into one or more of the following categories:

- **Species conservation** - where surveys may record species occurrence, distribution and population size.
- **Scientific research** - where survey data can add to understanding of bat behaviour and/or can be used to test scientific hypotheses.
- **Planning for development** - where a survey provides adequate information to enable:
    1. a planning authority to determine the effects of development on bat species and to identify and stipulate any further information required on necessary mitigation, compensation or enhancement measures; and/or
    2. an informed decision to be taken as to whether a Habitats Regulations licence should be applied for; and/or
    3. the relevant licensing body[6] to determine an application for a Habitats Regulations licence that would then enable the lawful disturbance of bats or the damage/destruction of their roosts.

Each of these categories is dealt with in more detail below.

2.2.1 Species conservation

Surveys may be required in order to inform and promote the conservation of bats in a specified area. Protocols for conservation surveys are the same as for other types of surveys in the same environment (e.g. surveys of buildings, trees, underground sites and open spaces).

6 Habitats Regulations licences are issued in England, Wales and Scotland under Regulation 44(2)(e) and in Northern Ireland under Regulation 39(2)(e) of the appropriate legislation for the purpose of *"preserving public health or public safety, or other imperative reasons of overriding public interest including those of a social or economic nature and beneficial consequences for the environment"*. Habitats Regulations licences are issued by the following bodies:
In England: Natural England.
In Northern Ireland: Environment and Heritage Service, Department of Environment Northern Ireland.
In Scotland: The Scottish Executive.
In Wales : The Welsh Assembly Government.

There are several types of survey that may be conducted for species conservation according to the objectives.

a) Presence/absence

This type of survey would record which species are present in an area; for example, to produce a county species atlas. Surveys should be designed to provide the necessary level of confidence in a negative record.

b) Usage

This type of survey would identify the species that use particular habitats or roost sites. This information could be used to inform conservation management and enhancement and/or to provide interpretation material for visitors to a conservation site; for example, to identify potential and actual tree roosts so that nature trails can be designed and located to avoid them. This would avoid conflict between the needs of bats and those of the public in locations where trees may need to be managed (e.g. through tree surgery) to ensure public safety. Surveys should be designed to provide the necessary evidence on which such decisions could be based.

c) Conservation management projects

A conservation licence issued by the relevant Statutory Nature Conservation Organisation (SNCO) is required for conservation projects that would otherwise constitute an offence, such as grilling the entrances of mines and caves or the alteration of roost sites. The proposed conservation activity may lead to the disturbance of bats or the damage of a roost site but there will be a net conservation benefit to bats at the site. A comprehensive survey would be required before a licence is issued and, as with development licences, a Method Statement detailing the proposed works and timetable would also be needed.

d) Surveys to inform potential site designation

A site being considered for statutory designation has to meet specific criteria, such as those given in the *Guidelines for the Selection of Biological SSSIs* (Joint Nature Conservation Committee, 1989). Survey work should be designed so as to provide the necessary information to establish if those criteria are met. The aim of the survey is usually to provide evidence that the site is used by a required number of bats in a given period of time.

Some Sites of Special Scientific Interest (SSSIs) designated because of bats consist of only a roost site (breeding roost or hibernaculum), but in some cases, key, non-roosting habitat has also been identified and included within the designated area. This is perhaps most relevant for sites that are also designated as Special Areas of Conservation (SACs), which can be protected for a species if the habitat *"presents the physical or biological factors essential to their life and reproduction"* under Article 4.1 of the Habitats Directive. Such factors include foraging and commuting habitat that enable bats to survive in the landscape in addition to roosting places.

e) Projects on or adjacent to a Special Area of Conservation (SAC)

All plans for projects that may affect SACs must (under Regulation 48 of the Habitats Regulations 1994 [43 of the Habitats Regulations (Northern Ireland) 1995]) be subjected to a screening process in order to determine if the proposal, alone or in combination with other proposals, will have a significant detrimental effect on the interests for which the SAC is designated. Detailed guidance on this process can be found in the document *Assessment of Plans and Projects Significantly Affecting Natura 2000 Sites – Methodological Guidance on the Provisions of Article 6 (3) and (4) of the Habitats Directive* (European Commission, 2001).

2.2.2 Research

Surveying for scientific research focuses on understanding bat lifestyles and behaviour. Studies are usually based on the testing of a hypothesis and the need to obtain objective evidence in support of it. The results and conclusions of such research can be important for conservation if they increase our ecological knowledge and lead to better informed conservation management decisions and the development of more effective mitigation. This in turn drives improvements in survey methods that surveyors concerned with conservation and development should adopt.

2.2.3 Planning for development

For development proposals requiring planning permission, the presence of bats, and therefore the need for a bat survey, is an important 'material planning consideration'[7]. Adequate surveys are therefore required to establish the presence or absence of bats, to enable a prediction of the likely impact of the proposed development on them and their breeding sites or resting places and, if necessary, to design mitigation and compensation. Similarly, adequate survey information must accompany an application for a Habitats Regulations licence required to ensure that a proposed development is able to proceed lawfully.

The term 'development' used in these guidelines includes all activities requiring consent under relevant planning legislation and/or demolition operations requiring building control approval under the Building Act 1984[8].

English Nature states that development in relation to bats *"covers a wide range of operations that have the potential to impact negatively on bats and bat populations. Typical examples would be the construction, modification, restoration or conversion of buildings and structures, as well as infrastructure, landfill or mineral extraction projects and demolition operations."* (Mitchell-Jones, 2004).

7 For instance – see paragraph 98 of ODPM Circular 06/2005 *Biodiversity and Geological Conservation – Statutory Obligations and their Impact within the Planning System.*

8 The Building Act 1984 is UK statute and is the enabling Act of the Building Regulations (as amended) in England and Wales. In Scotland, the Building (Scotland) Act 2003 is implemented by the Building (Scotland) Regulations 2004. In Northern Ireland, the Building Regulations (Northern Ireland) Order 1979 (as amended) is implemented by the Building Regulations (Northern Ireland) 2000 (as amended).

When considering development, competent authorities throughout the UK must have regard to the safeguarding of European Protected Species. Regulation 3(4) of the Habitats Regulations 1994 [also 3(4) in the Habitats Regulations (Northern Ireland) 1995] states: *"Without prejudice to the preceding provisions, every competent authority in the exercise of their functions, shall have regard to the requirements of the Habitats Directive so far as they may be affected by the exercise of their functions."*

This means that planning authorities must apply three tests when determining planning applications where European Protected Species may be involved. These tests mean they must have *"due regard"* to:

- the purpose of preserving public health or public safety or other imperative reasons of overriding public interest including those of a social or economic nature and beneficial consequences of primary importance for the environment - Regulation 44(2)(e) in the Habitats Regulations 1994 [39(2)(e) in the Habitats Regulations (Northern Ireland) 1995].

As long as:

- there is no satisfactory alternative – Regulation 44(3)(a) [39(3)(a)]; and

- the action authorised will not be detrimental to the maintenance of the species concerned at a Favourable Conservation Status in their natural range – Regulation 44(3)(b) [39(3)(b)].

These survey guidelines should therefore be used to inform surveys that are part of an application submitted for approval to a competent authority responsible for control and management of new development through the planning process.

Planning case law has established[9] that *"It is essential that the presence or otherwise of protected species, and the extent that they may be affected by the proposed development, is established before the planning permission is granted, otherwise all relevant material considerations may not have been addressed in making the decision[10]"*. Surveys should be carried out **before** planning permission is considered; only in exceptional circumstances should they be carried out **after** consent through planning conditions.

It is in the interests of all developers to ensure that sufficient information is submitted with a planning application to enable the competent planning authority to determine the application efficiently, lawfully and in a timely fashion. Survey data accompanying a planning application must be as up-to-date as possible. It is likely that the survey will provide only one part of the information that will need to be submitted. A full report will also include details of the proposed activity, an assessment of potential impacts and an avoidance, mitigation and compensation strategy[11] to deal with predicted adverse effects.

The presence of bats very rarely precludes development but planning conditions and obligations may be used to limit the extent of disturbance or restrict the timing of development activities. These may also stipulate mitigation, compensation and/or enhancement measures. Alternatively, or in addition, this may be achieved through the issue of a Habitats Regulations licence.

Government policy encourages developers to enter into pre-application discussions in order to seek advice on the information to be submitted with a planning application. Where bats are likely to be present (see Box 2.1), developers and consultants would be well advised to approach the planning authority at an early stage to discuss the need for a bat survey.

The implications of not carrying out an appropriate survey at a site that subsequently proves to have roosting bats can be significant. Insufficient information on protected species can render an application invalid (meaning that it will not be registered or determined) or lead to refusal or deferral of planning permission, resulting in significant delays and additional costs and may ultimately lead to a criminal prosecution if bats are harmed or their roosts destroyed.

Where a site has been designated as a Special Area of Conservation (SAC) because of its bat interest, the planning authority must *"make an appropriate assessment of the implications for the site in view of the site's conservation objectives"* - Regulation 48(1) [43(1)] **before** granting planning permission. It can only grant consent after it has ascertained that the proposal *"will not adversely affect the integrity of the European site"* - Regulation 48(5) [43(5)].

The guidelines may also enable Building Control Officers, in their role as a competent authority, to have regard to the Habitats Regulations[12] in relation to their functions under the Building Act 1984 where they are responsible for control and management of demolition operations[13] through the building control process.

9 Regina vs. The Cornwall County Council *ex parte* Hardy 2001 Journal of Planning Law 786.

10 This requirement has been strongly emphasised in government planning policy e.g. paragraph 99; ODPM Circular 06/2005 and in the Consultation Draft of Welsh Planning Technical Advice Note (TAN) 5 (20056).

11 A mitigation strategy should provide an assessment of the effects of the proposal and the options for avoiding, or reducing to an acceptable level, any negative impacts. Where it is not possible to avoid or reduce the effects of an impact to an acceptable level, the residual adverse impact may require compensatory measures to be implemented. This is usually written in the form of a Method Statement. Mitigation proposals are not dealt with specifically in these guidelines, but further information can be found in the *Bat Mitigation Guidelines* (Mitchell-Jones, 2004).

12 Where a Section 80 Notice (in England and Wales) has been served for the proposed demolition of a building, the presence of bats may be an important material consideration when a local authority's Building Control Section considers serving a Counter Notice (a Section 81 notice), setting out conditions that must be complied with during the course of demolition. In Scotland a building warrant is required for demolition (and all building work to which the Building (Scotland) Regulations 2004 apply).

13 When the owner of a building over $50m^3$ intends to have it demolished, the legal requirements of Sections 80-83 of the Building Act 1984 impose a requirement for formal notification to be given to the local authority at least six weeks before any demolition work starts. This is to enable all appropriate people and organisations to be advised of the proposed demolition and to ensure that measures to protect the public and/or to preserve public amenities are put in place.

2.3 Assessing the need for surveys

Every survey has a cost in time or money or both. It is therefore very important that the following are considered at the outset:

- that it is reasonable to request a survey;
- that the method (or suite of methods) is appropriate to meet the objectives of the survey; and
- that the survey effort is proportionate to the context and appropriate for the purpose of the survey.

The necessity of requesting a survey is considered in more detail below. Once it has been agreed that a survey is necessary, the appropriate methods and level of survey effort to employ must be decided during the preparation and planning process (see Section 3.8).

2.3.1 Necessity of requesting a survey

It is not necessary to have a record of bats at a location for a survey to be required. In practice, the locations of only a small minority of bat roosts have been formally recorded and even householders with large roosts of bats occupying their property may be unaware of their presence. Ephemeral (transitory) roosts such as those in trees are particularly difficult to detect.

The decision to undertake a bat survey must be based on a reasonable likelihood that bats may be present in the structure, tree, feature, site or area under consideration.

With regard to development, the government has stated that *"bearing in mind the delay and cost that may be involved, developers should not be required to undertake surveys for protected species unless there is a reasonable likelihood of the species being present and affected by the development*[14]*"*.

The trigger list in Box 2.1 gives common development situations where bats are *likely* to be encountered and therefore where it is most likely that a bat survey and assessment will need to be undertaken. It should be noted that the trigger list is a guide but it is by no means exhaustive. For example, seemingly unlikely habitats with few visible features for bats might provide good habitat at particular times of year. An example is moorland that has been found to have increased bat feeding activity during the heather flowering season.

Box 2.1 has been prepared in conjunction with the Association of Local Government Ecologists (ALGE), which in turn has been working with the Department of Environment, Food and Rural Affairs (Defra) and Natural England to prepare illustrative guidance[15] for local planning authorities, indicating when it is reasonable for surveys for protected species to be undertaken by developers. This is to ensure that the planning authority receives sufficient information with a planning application. During the scoping exercise that follows the early stages of a bat survey (desk study and walkover - see Sections 3.6 and 3.7), an assessment will be made of the types and levels of further survey, if any, that are appropriate and proportionate (see Section 3.8).

Development is recognised as having one of the most significant impacts on bats. The necessity of requiring surveys for other activities - such as forestry - is dealt with in other chapters - see, for example, Chapter 8.

The trigger list in Box 2.1 might also assist researchers and conservationists in focusing and prioritising their survey effort at particular sites or habitats with the most likelihood of finding bats.

14 Paragraph 99 ODPM Circular 06/2005 *Biodiversity and Geological Conservation – Statutory Obligations and their Impact within the Planning System.*

15 *Validation Checklists and Guidance Notes: Template for Biodiversity and Geological Conservation* (Association of Local Government Ecologists and others, in preparation). This is intended to act as a template for planning authorities who are encouraged to refine their own local requirements based on local knowledge and evidence about where bats are most commonly encountered within their area. Developers seeking advice on when a bat survey may reasonably be required should therefore also refer to any local protected species survey requirements specified by the relevant planning authority.

Box 2.1 Likelihood of bat presence – planning and development trigger list for bat surveys

Trigger list of where bats are *likely* to be present and where developers can reasonably be expected to submit a bat survey.

(i) Proposed development which includes the modification, conversion, demolition or removal of buildings and structures (especially roof voids) involving the following:
- all agricultural buildings (e.g. farmhouses and barns) particularly of traditional brick or stone construction and/or with exposed wooden beams greater than 20 cm thick;
- all buildings with weather boarding and/or hanging tiles that are within 200 m of woodland and/or water;
- pre-1960 detached buildings and structures within 200 m of woodland and/or water;
- pre-1914 buildings within 400 m of woodland and/or water;
- pre-1914 buildings with gable ends or slate roofs, regardless of location;
- all tunnels, mines, kilns, ice-houses, adits, military fortifications, air raid shelters, cellars and similar underground ducts and structures;
- all bridge structures, aqueducts and viaducts (especially over water and wet ground); and
- all developments affecting buildings, structures, trees or other features where bats are known to be present.

(ii) Proposals involving lighting of churches and listed buildings or floodlighting of green space within 50 m of woodland, water, field hedgerows or lines of trees with obvious connectivity to woodland or water.

(iii) Proposals affecting quarries with cliff faces with crevices, caves or swallets.

(iv) Proposals affecting or within 400 m of rivers, streams, canals, lakes, or within 200 m of ponds and other aquatic habitats.

(v) Proposals affecting woodland or field hedgerows and/or lines of trees with obvious connectivity to woodland or water bodies.

(vi) Proposed tree work (felling or lopping) and/or development affecting:
- old and veteran trees that are older than 100 years;
- trees with obvious holes, cracks or cavities; and
- trees with a girth greater than 1 m at chest height.

(vii) Proposed development affecting any feature or locations where bats are confirmed as being present, revealed by either a data trawl (for instance of the local biological records centre) or as notified to the developer by any competent authority (e.g. planning authority, Statutory Nature Conservation Organisation or other environmental or conservation organisation).

Notes

1 Bats may be found in other situations beyond those listed in Box 2.1. For example, pipistrelle bats will occupy modern buildings and built structures (see Chapter 6). Developers, and those acting for them, should be mindful that disturbance of any roosts or harm to a bat or bats is a criminal offence (see Section 1.5.1).

2 The trigger list in Box 2.1 is based on work done by the Durham Bat Group and Durham County Council and by the Surrey Bat Group and Surrey County Council. The criteria also draw on the *Bat Mitigation Guidelines* (Mitchell-Jones, 2004).

3 The trigger list in Box 2.1 will be particularly useful in signalling a bat survey in situations where a population of a species might be present and an action would consequently be detrimental to the maintenance of a population of a species at a Favourable Conservation Status and therefore would be contrary to the statutory requirements of Regulation 44(3)(b) of the Habitats Regulations 1994 [39(3)(b)] (see Section 2.2.3).

4 Where bat roosts are of international importance, they may have been designated as Special Areas of Conservation (SACs). There are cases where larger scale developments 5-10 km away from such sites have been required to undertake bat surveys and impact assessments in order to account for foraging bats using extensive areas around their roost.

Chapter 3

Preparation and planning

3.1 Introduction

Once it has been decided that a survey is required, preparation and planning will be needed. This chapter provides guidance on this process and covers the skills and experience necessary, licensing and health and safety requirements, preliminary data gathering by desk study and site walkover, the selection of appropriate methods and timings for further survey, and planning for the reporting and recording of surveys.

3.2 Skills

Surveying for bats is a highly skilled task requiring training and should only be carried out by those with relevant experience. Anyone commissioning surveys or reviewing them should be sure that the surveyor has the necessary expertise. It is the duty of all surveyors to ensure that they are experienced in, and trained for, the work that they are undertaking. Surveyors should be aware of their own limitations in terms of knowledge and experience; for example, many bat workers outside Wales and south-west England will have little or no experience of working with horseshoe bats. This should be taken into account if asked to undertake work in these parts of the UK.

While membership of a professional institute such as the Institute of Ecology and Environmental Management (IEEM) or a Chartered Environmentalist (CEnv) qualification does not test a skill level with respect to bats or other species, the conditions of membership do require members to undertake surveys in a professional manner and make clear any limitations to their work. Currently in the UK, there is no agreed standard for bat survey work to which ecologists or zoologists should operate, and no specific qualification that should be gained. The Bat Conservation Trust is in the process of developing standards in bat training which will outline the knowledge and skills required for proficiency in different tasks relating to bat work.

The Statutory Nature Conservation Organisations (SNCOs) license bat workers for conservation, science or education. Potential workers must undergo training and peer review before being licensed as a bat worker. The possession of such a licence is an indication that the surveyor has reached a minimum standard of training, although this may not relate to survey work, impact assessment or the design and implementation of mitigation schemes. It is reasonable for those hiring bat consultants to ask for evidence of previous work, in order to demonstrate the surveyor's competence in the type of work they are being requested to undertake. By the same token, those surveying and reporting on surveys, especially unlicensed bat workers, should include evidence of their competence in the documents they produce.

Basic skills needed for bat surveying can often be gained initially through membership of a local bat group or from attendance at a bat foundation course where a basic knowledge of bat biology and behaviour is acquired. Thereafter, attributes required include a good ear to hear differences in sounds produced through heterodyne bat detectors, sharp eyesight to find droppings, urine staining etc., good vision in poor light levels, agility to gain access to potential bat roosting sites and stamina to remain awake and alert during late night or early morning survey sessions! Not only is it important to know what to look for but also where to look for it, as this in itself can be diagnostic of the species present. Key skills to be acquired are an understanding of building, barn, bridge and tree structure and terminology, and field signs including bat remains, droppings and their identification. Also, a thorough knowledge of species ecology, with particular reference to roosting ecology and therefore likely location of bats (especially when out of sight), and appropriate mitigation is also required.

3.3 Training

A thorough knowledge of bat biology, including keeping abreast of latest research findings, is necessary in order to undertake surveys and interpret the findings; to do this, regular formal or informal training is necessary.

Training in bat surveying can be sourced in a number of ways. Many consultants learn in-house through mentoring and/or attending a variety of training courses offered by a number of individuals and organisations, including the Bat Conservation Trust (BCT). These range from courses on basic surveys using bat detectors to more specialised courses on particular aspects of survey, for example, surveying buildings, barns or trees. Other courses that may be of relevance include those on planning for protected species. Some bat groups may allow consultants to gain experience with them, especially if voluntary commitment is made to the group in return. BCT's National Bat Conference aims to update bat workers with the latest research findings through its speakers and workshops.

For further information about training courses offered by the BCT, request a current training brochure from the BCT National Bat Helpline (0845 1300 228) or download a copy from the website **www.bats.org.uk**.

3.4 Licensing for bat survey

This section provides a brief overview of the system for licensing bat surveyors. For detailed information, the relevant licensing body and its guidance should be consulted (see also Section 1.5.1).

A surveyor may need to undertake activities during survey work that would violate the strict protection afforded to bats by the Habitats Regulations. A licence is required to permit such activities, for example, entry into a bat roost and temporary disturbance of bats during the survey or capture and marking with radio-transmitters.

These licences are varyingly referred to as SNCO bat licences, science/education licences, survey licences or conservation licences. These titles are reflective of the purposes for which they are granted under the Habitats Regulations and the fact that the SNCOs regulate the issue of such licences (see also Section 1.5.1). These are personal licences allowing a surveyor to disturb, handle or mark bats, or to improve their roost sites where the main purpose of the work is for conservation of the species at that site. They do not cover work such as destruction of a roost site for purposes of development, overriding public interest or public health and safety, which would be covered by a Habitats Regulations licence (see also Sections 1.5.1 and 2.2).

It is best practice for surveys of potential roosts to be done by licensed surveyors because if a roost is discovered and needs to be entered, an SNCO licence is required as bats may be deliberately disturbed. On discovering a roost in a previously unknown location, an unlicensed bat worker must withdraw.

Only an SNCO licensed bat worker can enter a known roost or area occupied by bats. However, it is not always necessary for surveyors to have an SNCO bat licence. For example, a licence is not needed for undertaking activity surveys using bat detectors in the field or emergence surveys outside roosts as these do not cause disturbance to bats when undertaken properly.

Licences for more invasive survey methods, such as mist netting, marking and radio-tracking, are usually issued for specific projects requiring their use. The licences are issued for a set period of time, at a particular location and for a given number of animals. The applicant will need to demonstrate that the level of disturbance is justified and that they have the necessary experience and training to undertake the work.

3.5 Health and safety

Health and safety is an important issue for anyone organising a bat survey whether for development, research, conservation or enjoyment.

Bat surveys can be difficult, involve challenging locations, unsociable working hours, extensive travel, disrupted sleep and many potential hazards. It is important that these risks are adequately considered and accounted for during the planning of each survey. While bat conservation is important, it should be remembered that the most important mammals on any survey are the surveyors.

3.5.1 Health and safety policy

It is the legal duty of an employer to have a health and safety policy. Section 2(3) of the Health and Safety at Work etc. Act 1974 (HSW Act) states: *"Except in such cases as may be prescribed, it shall be the duty of every employer to prepare and as often as it may be appropriate revise a written statement of his general policy with respect to the health and safety at work of his employees and the organisation and arrangements for the time being in force for carrying out that policy, and to bring the statement and any revisions of it to the notice of all his employees"*.

While employers with fewer than five employees are not required to have a written health and safety policy, they are still liable for their health and safety and should consider it best practice to have and uphold such a policy.

The *Working at Height Regulations 2005* cover all instances where a task involves working at height and there is a risk of falling. Although the obvious scenarios for working at height during bat surveys include work in buildings, on structures such as bridges and in trees, working at height may also arise while surveying underground or on quarry faces. The *Working at Height Regulations 2005* make particular reference to the use and limitations of ladders as a means of access. For further information, these and other Regulations can be downloaded from the Office of Public Sector Information website **www.opsi.gov.uk**

Due to the unsociable hours often inherent in bat work, surveyor tiredness is a very real risk. *The Working Time Regulations 1998* (as amended) advise a *"rest period of not less than eleven consecutive hours in each 24-hour period"*.

Where relevant health and safety regulations and guidance exist, it is good practice to adhere to them. Guidance on safety and risk management-related issues can be found on the Health and Safety Executive's website **(www.hse.gov.uk)** or by calling the HSE Infoline (0845 345 0055).

Table 3.1 Hazards and risks associated with fieldwork and procedures and equipment used in management of those risks.

Risks particularly associated with manual bat activity surveys in the field[a], with roost surveys of buildings[b], trees[c] and underground sites[d] are marked accordingly.

Types of hazards and risks associated with fieldwork	Procedures to remove or reduce risk posed by the hazard	Equipment to remove or reduce risk posed by the hazard
Lone working	Lone working should be avoided wherever possible. If it is unavoidable, a buddy system and late working procedure should be put in place. It is particularly important for a second person to be present when working at height. Always park your car so that you can drive straight away rather than have to do a three-point turn. This is useful in case of emergencies but also in cases of aggression.	In case of separation or accident a mobile phone (or satellite phone in remote areas), two-way radio, whistle, map and compass should be carried.
Tiredness[a, b, c, d]	Limits to the number of surveys being carried out are recommended: • Staff doing bat surveys only (with no daytime work) – up to four dusk or four dawn or two of each or two all-nights per week. • Staff doing other work – up to two dusk or two dawn or one of each per week. All-night surveys are not appropriate.	
Bad weather[a, c]	Awareness of the weather forecast.	All-weather clothing appropriate to the local situation (fieldwork in the north of Scotland will require a greater level of protection than in the south of England at the same time of year).
Work in the dark[a, b, d]		Powerful torch (with spare torch, batteries and bulbs).
Work in confined spaces[b, d]	Confined spaces training – see Section 3.5.3.	
Work underground where there may be sudden drops, changes in roof height, unstable rock or decaying fixtures[d]	Mine safety training – see Section 3.5.3.	Protective warm clothing, strong boots, helmet and helmet-mounted lamp. Ladders and/or ropes if necessary for the site.
Work at height e.g. in attics[b], trees[c] or on quarry faces where there is a risk of falling[d]	Arboricultural climbing training – see Section 3.5.3	Safe means of access e.g. ladders, platforms and ropes.
Busy roads, on railways or on farmland with working agricultural machinery[a, b]		Fluorescent jacket (appropriate to road type).
Derelict structures/ construction sites[b]/trees[c] where there is risk of falling masonry or tree branches		Hard hat, fluorescent jacket, safety footwear.
Water (rivers, streams, ditches, lakes, canals etc.)[a]	Employ safe methods of crossing watercourses such as rivers, streams and ditches. Check the flood conditions.	Life jacket (consider self-inflating type to allow for greater mobility).
Unfenced slurry or silage pits, ponds, grain silos and stores[a]	Take care in vicinity of such areas.	

Slips, trips and falls caused by rough ground[a, b, c, d]	Take care in moving around, ensure visibility is adequate.	
Sunburn/sunstroke	Use sunscreen, hat, long-sleeved shirt.	
Diseases such as Weil's disease, Lyme disease, ornithosis (an infectious disease that affects birds and can affect humans and other mammals) and tetanus (for example, from rusty barbed wire)	Awareness of diseases e.g. carrying a Weil's disease awareness medical card or being familiar with tick identification.	Protective clothing with any open cuts covered. Ornithosis – protective dust mask and gloves.
Insect bites and stings (horseflies, ticks, etc.)		Insect repellent.
Bat bite[b] and rabies (European Bat Lyssavirus)	All those who handle bats should be vaccinated against rabies because of the risk of European Bat Lyssavirus. Care should be taken when handling bats to avoid getting bitten. Information and advice on vaccinations for bat workers and what to do in the event of a bat bite can be obtained from the Health Protection Agency's website **(www.hpa.org.uk)** or by calling the HPA Centre for Infections (020 8200 4400). Further information on vaccination is available in the Department of Health 'Green Book' *Immunisation Against Infectious Disease 2006* which can be obtained from the Department of Health's website **(www.dh.gov.uk/en/Policyandguidance/Healthandsocialcaretopics/Greenbook/index.htm).**	Appropriate gloves should be worn when handling bats (advice is available from the Bat Conservation Trust).
Asbestos, fibreglass and dust[b]	Every non-residential building should have an Asbestos Register. Ask to see it for the building to be surveyed, particularly those built between 1950 and 1985. Asbestos should be avoided and specialists called. See Section 3.5.3 for details of asbestos training.	Asbestos - disposable overalls and respirator. Fibreglass and dust – protective dust masks (conforming to BS EN149), safety glasses and overalls.
Sharp objects, such as broken glass or hypodermic syringes[a, b]		Safety work boots with protective toecaps and reinforced soles, impact grade gloves, overalls, first aid kit.
Land that has been sprayed[a]	Check with landowners/agents to establish whether pesticides have recently been used on land being accessed. Many pesticides have a 'harvest interval' between spraying and harvesting and this is also the period when people should not enter the land.	
Aggressive farm animals such as guard dogs, geese, bulls and cows with calves at foot[a, b]	Check with landowners/agents to establish the locations of livestock, guard dogs, geese, etc.	
Shooting/predator control (often takes place at dusk)[a]	Check with landowners/agents whether any predator control is likely to be taking place.	
Verbal and physical assault[a]	Avoid lone working, work within sight of accompanying surveyor, park so as to be able to leave quickly.	Attack alarm.

3.5.2 Hazards, risks and risk management

All jobs carry risks. Bat surveying and fieldwork have some very specific risks from particular hazards. A **hazard** has the potential to cause harm; it is associated with degrees of danger and is quantifiable. **Risk** is the likelihood of harm from a particular hazard and its severity i.e. risk = severity x likelihood. If the risk is too high the action should not be undertaken.

The simplest way to ensure protection from harm is to remove the hazard or the risk that the hazard poses. Use common sense – if it seems dangerous, do not do it unless you can make it safe first! No activity should be attempted without undertaking measures to bring the likelihood of risk within acceptable limits. Many of the risks associated with bat survey and fieldwork can be avoided, removed or reduced by employing measures put in place as a result of thorough risk assessment. Table 3.1 lists risks associated with bat fieldwork and measures that can be taken or equipment that can be used to manage that risk.

There are cases of surveyors/handlers suffering very serious consequences from their work. Examples include:

- a bat surveyor/handler who died as a result of being bitten by a bat suffering from European Bat Lyssavirus 2;
- a surveyor shot in the chest by a gamekeeper who was 'lamping' foxes;
- surveyors attacked and their equipment stolen while undertaking a bat survey on the edge of an urban area; and
- car accidents in which a contributory factor may have been excessive tiredness following night-time surveys.

A risk assessment may take many forms but it is principally a list of identified potential site-specific hazards and the means used to reduce or remove them. Relevant facts should be formally set down in a risk assessment form which complies with legal requirements as well as being best practice.

A risk assessment should be prepared and completed for every job undertaken. Targeted risk assessments will be required for every site, to ensure risks in addition to those listed in Table 3.1 are taken into account. On arrival on site, the risk assessment should be reviewed to establish that all possible risks have been taken into account.

Guidance on risk assessments for bat and other survey work is available from various sources including the *Bat Workers' Manual* (Mitchell-Jones and McLeish, 2004), the Institute of Ecology and Environmental Management (available only to members), and from sample risk assessments, for example, from volunteer schemes such as the BCT's National Bat Monitoring Programme.

In some complex situations, such as where a building is unsafe and asbestos may be present, a Method Statement may also be needed.

As most bat survey work involves at least one, or a combination of several, risk factors it will seldom be appropriate for one person to undertake a survey unless accompanied on site.

All equipment used should be regularly checked and maintained, in line with appropriate legislation.

3.5.3 Health and safety training

Training courses are available for those likely to visit high-risk sites or to undertake high-risk activities during surveys. High-risk sites include:

- construction sites;
- roads;
- railways;
- enclosed spaces;
- caves, quarries and mines;
- derelict buildings;
- buildings and structures which contain asbestos;
- water bodies;
- treatment works;
- industrial plants;
- remote locations; and
- sites with criminal activities and/or hostile local residents.

The following locations require advanced knowledge and the use of specialist equipment, practice in the use of which can be gained on the specialist training courses indicated:

- work in confined spaces (tunnels, culverts etc.) - Confined Spaces training course;
- work in trees - Arboricultural Climbing course with specialist equipment;
- work underground (mines, caves etc.) - Confined Spaces training course or Mine Safety course. Basic caver training and advice on safety issues in specific local caves and mines can be obtained via the British Caving Association (BCA), Regional Caving Councils or local caving clubs.

Following a number of fatalities, the Health and Safety Executive has introduced a certification scheme for all those who need to work on active construction sites. A course is available from the Construction Industries Training Board (CITB) to obtain a Construction Site Certification Scheme (CSCS) card. These are now often requested before entering a construction site.

It is advisable to consider attending an Asbestos Awareness training course if entering buildings with asbestos could become a part of survey work.

3.6 Desk study

A desk study or data trawl should always be conducted at an early stage when planning bat surveys. It is a prerequisite to deciding the type and intensity of further survey where required. However, it is unlikely that a desk study alone would ever provide enough information to fully assess the value of an area for bats; an initial visit to the site for a walkover or scoping survey is recommended (see Section 3.7). Together, the desk study and walkover survey should inform the scoping exercise that determines what further survey work is required (see Section 3.8).

3.6.1 Information sources

The first step in any data trawl exercise should be to obtain maps of the area. These can be supplemented with aerial images, which may be obtained from sources such as Google Earth (**www.earth.google.com**) and Multimap (**www.multimap.com**). From such maps, an experienced surveyor can gain an impression of the habitats and features likely to be important for bats and make a judgement of the species likely to be present and determine where best to look for them.

Distribution atlases, such as the *Distribution Atlas of Bats in Britain and Ireland* (Richardson, 2000), should be consulted. Local distribution atlases should also be consulted where available. The date of publication of such sources should be borne in mind; more recent records may exist but not be represented.

Online sources of distribution data include the National Biodiversity Network (**www.searchnbn.net**); however, the likelihood is that not all records exist on such databases and not all parts of the UK have been surveyed so the data coverage may be patchy. The absence of a species record in an area does not necessarily mean that the species is not there.

It is important to obtain known information about bat roost sites or any sites of nature conservation importance designated for their bat interest near to, or on, the site in question. When requesting information, it is necessary to provide a 6-figure grid reference for the site, the radius within which searches are requested and to state the species, age of data and type of record required.

There are a number of information sources for bat roost records and sightings, including:

- Local offices of the Statutory Nature Conservation Organisations (SNCOs). A valuable source of information, especially as they receive bat roost report forms from their network of voluntary bat wardens. SNCOs should be contacted at the start of a project to obtain any relevant data. However in many cases, offices do not hold these records in an accessible format.

- Local Biological Record Centres (known as LRCs or BRCs). These are found in most counties and will undertake a data trawl of their records for a fee. A list of active LRCs can be found on the National Federation for Biological Recording (NFBR) website (**www.nfbr.org.uk**).

- A Biodiversity or Nature Conservation Officer (also known as county ecologists), who may have access to records, is employed by some local, county or district councils.

- Local bat groups usually hold a database of bat records. The Secretary should be able to provide the name and contact details of the member who maintains the database. The local bat group contact details can be obtained from the BCT website (**www.bats.org.uk**) or by calling the BCT National Bat Helpline (0845 1300 228).

- Local wildlife trusts also keep bat records (see **www.wildlifetrusts.org**). Sometimes the local bat group is part of the local wildlife trust, so the information can be from the same source.

- County mammal recorders. These are volunteer recorders who collate records sent to them about mammal sightings in their county. Contact details are available from the Mammal Society website (**www.abdn.ac.uk/mammal/**).

- Local or national mining history or caving groups and clubs and caving councils may have useful information. See the British Caving Association (**http://british-caving.org.uk/**) for details. A number of cave systems have biological recorders and records are often published in club or regional journals.

- On-site personnel such as site security guards, caretakers or gardeners. They may provide anecdotal evidence that gives useful pointers but this may not be robust or reliable enough to contribute to a desk study.

The study site may be on the boundary of two or more organisations, in which case they should all be contacted.

When using or referring to materials obtained from external sources, rules of copyright should be noted and adhered to. There may also be restrictions on the commercial use of internet resources.

3.6.2 Geographical extent of desk study

It is important to request information up to at least 1 km from the survey site. Depending on the nature of the proposed project, its scale and the species likely to be affected, a larger radius search area may be required (see also Chapter 2). If there are, for example, Special Areas of Conservation (SACs) or other designated sites (e.g. Sites of Special Scientific Interest - SSSIs) within the radius selected for data searches, or if the extent and potential impact of the proposed works is especially high, information should be requested up to 10 km away from the site. In exceptional circumstances, such as where rare species that travel further than 10 km may be present, data trawls may need to be requested over larger areas.

Where the proposed work may affect a SAC, an Appropriate Assessment may be required (see also Section 2.2.3).

3.7 Site walkover survey

It is unlikely that a desk study or data trawl (see Section 3.6) alone would ever provide enough information to fully assess the value of an area for bats; therefore, an initial visit to the site for a walkover survey is strongly recommended. Site walkovers are sometimes referred to as 'scoping surveys'. They should aim to identify the potential value of the habitat for bats. Information gained from being on-site is a prerequisite to designing the type and intensity of further survey, should it be required. In this way the walkover or scoping survey contributes important data to the 'scoping exercise' (see Section 3.8). It is an advantage to undertake the site walkover with knowledge of the desk study.

In larger projects, a walkover would normally be part of an initial ecological appraisal. With the necessary awareness, a professional ecologist can look for features of potential value to bats at the same time as recording other features of ecological value. If bat or bird boxes are present, they should be checked for use by bats either at this stage or during a subsequent survey (note that surveyor licences are required for bat box checking). Small sites may not necessarily require a specific site walkover for bats, as this assessment could be done in conjunction with a Phase I Habitat Survey[16].

All site visits require a health and safety risk assessment and appropriate permission for site access. The visit should also be used to look for, and take note of, additional potential health and safety hazards (see Section 3.5).

Surveyors undertaking the site visit are responsible for identifying and recording areas and structures of potential value for bats. The survey is usually undertaken by walking over a site and observing the features present. It should be possible to identify those features most likely to be of value for bats and these should be marked on a map or plan. The walkover survey should record the value of each feature on site or in the landscape according to its potential for use by bats for roosting, foraging or commuting, taking into account its quality.

Guidance on assessing the value for bats of habitat features within the landscape is given in Box 3.1. A continuum is presented between low and high potential. Individuals may wish to assign habitat as low, medium or high value for bats. For example, in a river valley the following features would be identified as having high value to bats and therefore indicative of high likelihood of bat presence:

- older trees/woodlands for foraging and roosting;
- linear landscape elements e.g. hedgerows and watercourses for commuting and foraging; and
- built structures e.g. buildings and bridges for summer roosting or hibernation.

Alternatively, features could be assigned a level of bat potential when they are examined in more detail (by climbing/internal inspection/emergence – see relevant later chapters).

16 Phase I Habitat Survey is a standardised system for surveying, classifying and mapping wildlife habitats including urban areas. For further information see *Handbook for Phase 1 Habitat Survey - A Technique for Environmental Audit* (Joint Nature Conservation Committee, 2004) and the *National Vegetation Classification: Users' Handbook* (Joint Nature Conservation Committee, 2006).

Box 3.1 Guidance for assessing the value of habitat features within the landscape for bats and hence the likelihood of bats being present.

Note
There are no clearly defined categories of habitat value; rather there is a continuum from low to high value for bats. Expert judgement will be required when assessing the relative value of a site for bats based on the features identified and the context in which the site or survey area is located.

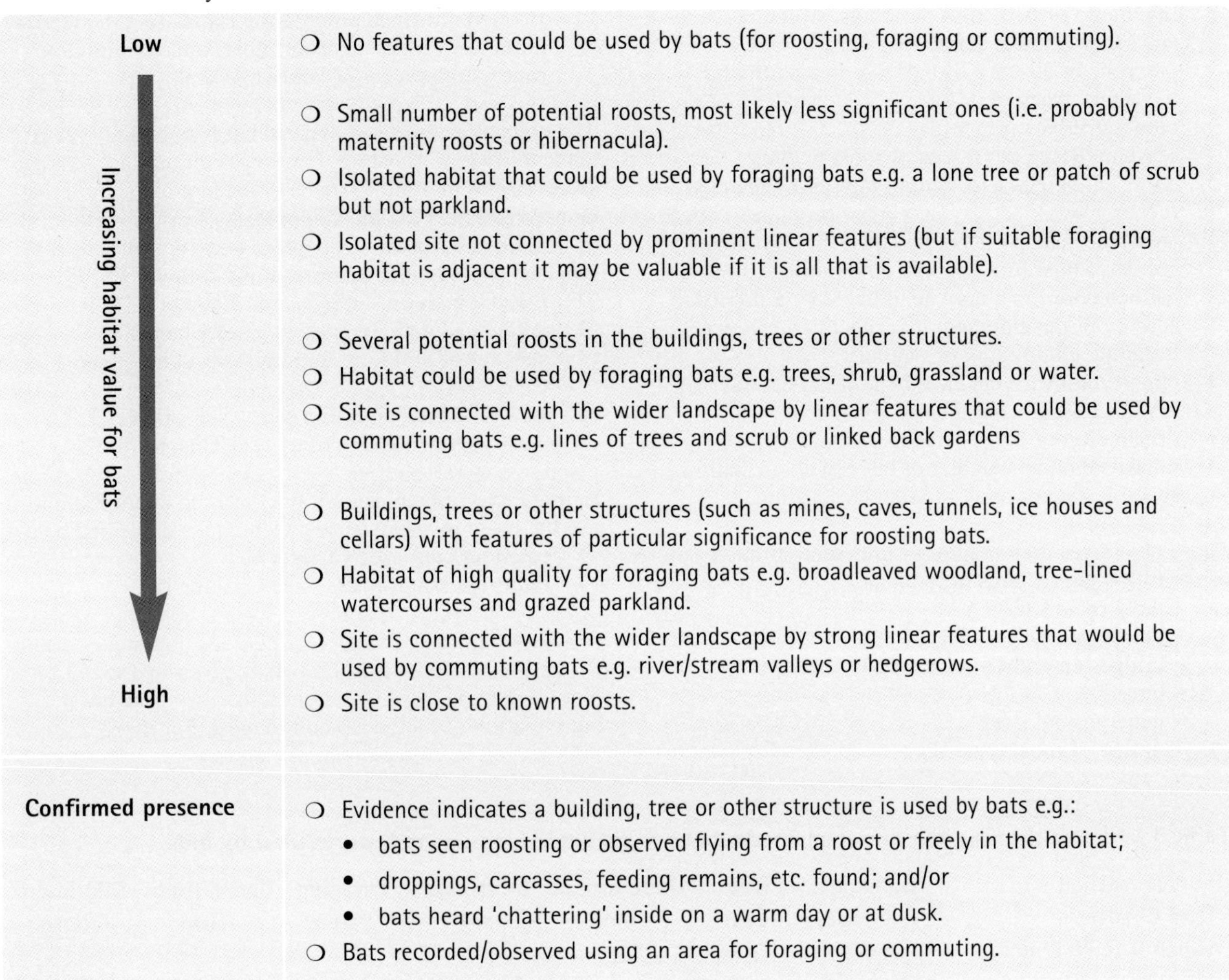

Increasing habitat value for bats	
Low	❍ No features that could be used by bats (for roosting, foraging or commuting).
	❍ Small number of potential roosts, most likely less significant ones (i.e. probably not maternity roosts or hibernacula). ❍ Isolated habitat that could be used by foraging bats e.g. a lone tree or patch of scrub but not parkland. ❍ Isolated site not connected by prominent linear features (but if suitable foraging habitat is adjacent it may be valuable if it is all that is available).
	❍ Several potential roosts in the buildings, trees or other structures. ❍ Habitat could be used by foraging bats e.g. trees, shrub, grassland or water. ❍ Site is connected with the wider landscape by linear features that could be used by commuting bats e.g. lines of trees and scrub or linked back gardens
High	❍ Buildings, trees or other structures (such as mines, caves, tunnels, ice houses and cellars) with features of particular significance for roosting bats. ❍ Habitat of high quality for foraging bats e.g. broadleaved woodland, tree-lined watercourses and grazed parkland. ❍ Site is connected with the wider landscape by strong linear features that would be used by commuting bats e.g. river/stream valleys or hedgerows. ❍ Site is close to known roosts.
Confirmed presence	❍ Evidence indicates a building, tree or other structure is used by bats e.g.: • bats seen roosting or observed flying from a roost or freely in the habitat; • droppings, carcasses, feeding remains, etc. found; and/or • bats heard 'chattering' inside on a warm day or at dusk. ❍ Bats recorded/observed using an area for foraging or commuting.

The walkover survey should be undertaken in daylight hours, with sufficient time allowed to walk the whole area.

Consideration should be given to whether the walkover area should be greater than the development footprint (impacts, for example, from lighting, could extend beyond the site boundary) and extensive sites may require more than one visit for full coverage.

3.8 Selection of survey methods

The information gathered from the data trawl and walkover should be collated and assessed in relation to the proposed development and used to inform the type and extent of future bat survey work. This is essentially a 'scoping exercise' that identifies the possible effects of the development on bats and sets the future survey objectives. It should include consideration of appropriate survey methods and the level of survey effort and area coverage that is proportionate.

As with the desk study and walkover survey, it is important to consider the area that the scoping exercise should cover. This could simply be the scheme footprint for very small-scale proposals; however, for larger developments, such as a new highway or a planned residential development, it should cover an area that takes into account the wider landscape. When deciding the extent of any assessment, it is important to consider the species potentially affected and the nature of the proposed works.

3.8.1. Appropriateness of survey method

Once it has been established that further survey is required beyond that conducted in the desk study and walkover survey, the next step is to establish the most appropriate survey in order to obtain adequate results for the intended purpose.

The purpose of the survey should be defined through clear objectives (see Section 4.1.2) and these should guide the selection of appropriate survey methods and the amount of survey effort required (see Section 3.8.2).

For instance, the objectives of the survey could be to establish one or more of the following:

- likelihood of particular buildings, structures, trees and other features supporting bats;
- presence or absence of bats e.g. in a particular building, structure or tree;
- number of bats present;
- seasonal usage of an area or roost by bats;
- specific features used within the survey area by roosting bats;
- areas and features of importance for particular species whilst foraging and commuting;
- bat behaviour that may be affected by a proposed activity or development in terms of emergence, foraging, commuting or mating;
- opportunities for enhancement of bat habitat that may be possible within any given area;
- distribution of bat species in a landscape;
- improved understanding of bat behaviour within a specific species or at a particular location.

These objectives may require examination of one or more of the following features or habitats:

- Likely roost sites e.g.:
  - buildings;
  - bridges and other structures;
  - trees; or
  - underground sites.
- Confirmed roost types e.g.:
  - summer maternity roosts;
  - winter hibernacula;
  - mating roosts;
  - swarming sites; or
  - other/transitory roosts.
- Likely/confirmed foraging habitat (e.g. woodland, grassland, riparian, heathland, marsh/bog or other).
- Likely/confirmed commuting routes (e.g. hedgerows, treelines, watercourses or other) and/or migration routes at the local or landscape scale.

The objectives, location, type of bat habitat and type of bat activity (e.g. roosting, foraging, commuting, mating, swarming or migrating) under consideration should then be used to select the most appropriate survey method(s) as described in detail in Chapters 4 to 10. Methods may typically involve one or more of the following:

- manual activity surveys - see Chapter 4;
- automated activity surveys - see Chapter 5;
- surveys of buildings and built structures - see Chapter 6;
- surveys of underground sites - see Chapter 7;
- surveys for roosts in trees - see Chapter 8.

In more exceptional cases it may also be appropriate to use the more invasive survey techniques described below:

- catching bats - see Chapter 9;
- radio-tracking of bats - see Chapter 10.

A matrix depicting the survey methods that are available for various features used by bats is given in Table 3.2. Not all methods will be required in every case but a combination of methods is often used. For more information see the individual chapters.

Table 3.2 Matrix showing the survey methods that are available for various features used by bats

Survey method (and Chapter in which it is described)	Building or bridge roost	Tree roost	Underground roost	Swarming site	Foraging area	Commuting route	Migration route
Non-invasive methods							
Internal inspection survey (Chapters 6, 7 & 8)	✓	✓	✓				
External inspection survey (Chapters 6 & 8)	✓	✓					
Emergence/re-entry surveys (Chapters 4, 6, 7 & 8)	✓	✓	✓	✓			
Backtracking (Chapter 4 & 8)	✓	✓					
Manual bat activity surveys (Chapter 4)					✓	✓	✓
Automated bat activity surveys (Chapter 5)	✓		✓	✓	✓	✓	✓
Invasive methods							
Catching surveys (Chapter 9)	✓	✓	✓	✓	✓	✓	✓
Radio-tracking surveys (Chapter 10)		✓			✓	✓	✓

3.8.2. Cautionary note - non-invasive versus invasive techniques

Resolution 4.6 of the EUROBATS Agreement[17] gives guidance on invasive survey techniques such as for catching bats. It states that *"the research being proposed should not adversely affect the conservation status of the population and should take account of the welfare of individual bats"*. It also states that radio-tracking *"should only be used for well-organised and authorised projects where essential data cannot be acquired with less intrusive methods"*. Therefore when surveys are undertaken for development, non-invasive survey methodologies should preferably be exhausted first before invasive techniques are employed; where they are used, constant re-appraisal of the value of the information in relation to the potential risks to the bats is essential. Disturbance caused by a survey should be the minimum required to obtain the necessary information and the least intrusive methods possible should always be employed.

3.8.3. Cautionary note - proving absence

It is comparatively easy to determine use of a site by bats, but absence is more difficult to prove. It requires greater effort to demonstrate beyond reasonable doubt that bats are not present or likely to be present.

3.8.4. Proportionality of survey

The type of survey undertaken and the amount of effort expended needs to be proportional to:

- the type and scale of the proposed activity or project and its predicted impacts on bats;
- the likelihood of bats being present or affected;
- the species and numbers of individuals concerned; and
- the type of roost and/or habitat affected.

Greater effort should be expended in situations where high numbers of bats are likely to be present and where there is a high degree of risk that they may be adversely affected. For example, the development of a large housing scheme is likely to have a range of effects and therefore more detailed surveys at a landscape scale are likely to be appropriate. The amount of survey effort will also depend on the geographic region, the species found during the data trawl and the type of roosting and foraging habitat in the locality. If these factors suggest the likelihood of an Annex II Habitats Directive species being present (one for which a SAC may be designated) then a greater level of survey effort is likely to be required.

Figure 3.1 is a flow chart depicting the decision-making process and gives guidance as to the level of survey that is likely to be proportional according to the likelihood of bats being present. It is worth noting that the type of survey to be undertaken and amount of effort expended can often only be fully determined after visiting the site at least once (see Section 3.7). Surveying for bats is often an iterative process, with later surveys being informed by earlier ones. It may often not be possible to determine all survey requirements at the start of a project.

A survey will only provide a snapshot of activity at the time it is undertaken. Additional surveys may be needed where information on temporal or seasonal changes in activity are needed, where a significant period of time has elapsed since the previous survey or where surveys have been limited by seasonal, weather, access or other constraints. An ecologist experienced in bat work should be able to recommend when further surveys are required. If the survey and management or mitigation proposals have been fully able to take into account potential use of a site during other seasons, there is little merit in delaying works, provided there is little chance of further surveys finding any new significant facts.

17 EUROBATS Agreement Resolution 4.6 *Guidelines for the Issue of Permits for the Capture and Study of Captured Wild Bats* is contained within the written record of the Fourth Meeting of the Parties to the Agreement. Resolution 4.6 was subsequently amended by Resolution 5.5 at the Fifth Meeting of the Parties to the Agreement. Both meeting records are available online from **www.eurobats.org**. The UK is a Party to the Agreement.

Figure 3.1 Flow chart depicting the process of deciding what level of survey is necessary

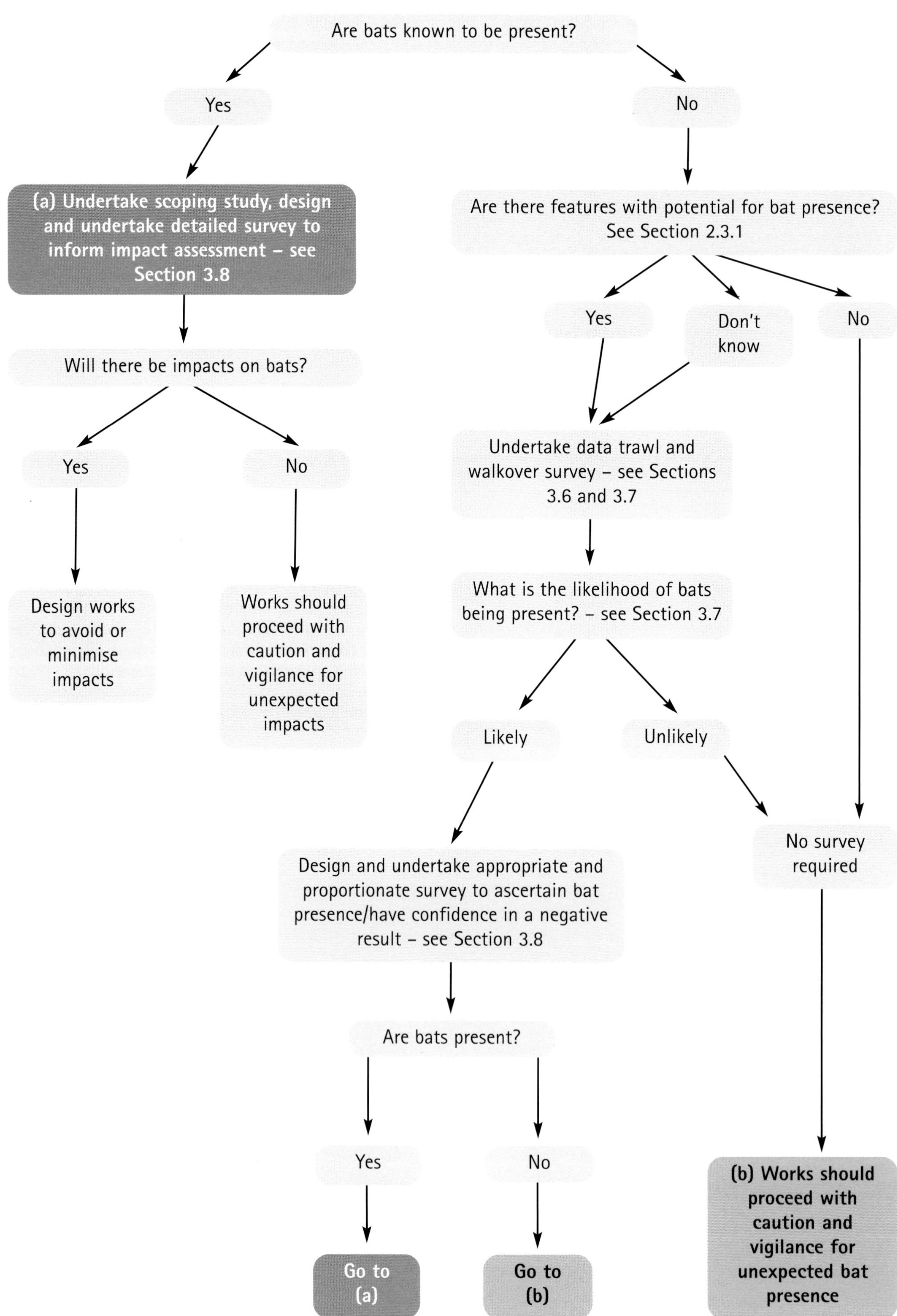

3.9 Survey timings

Figure 3.2 Overview of a year in the life of a bat

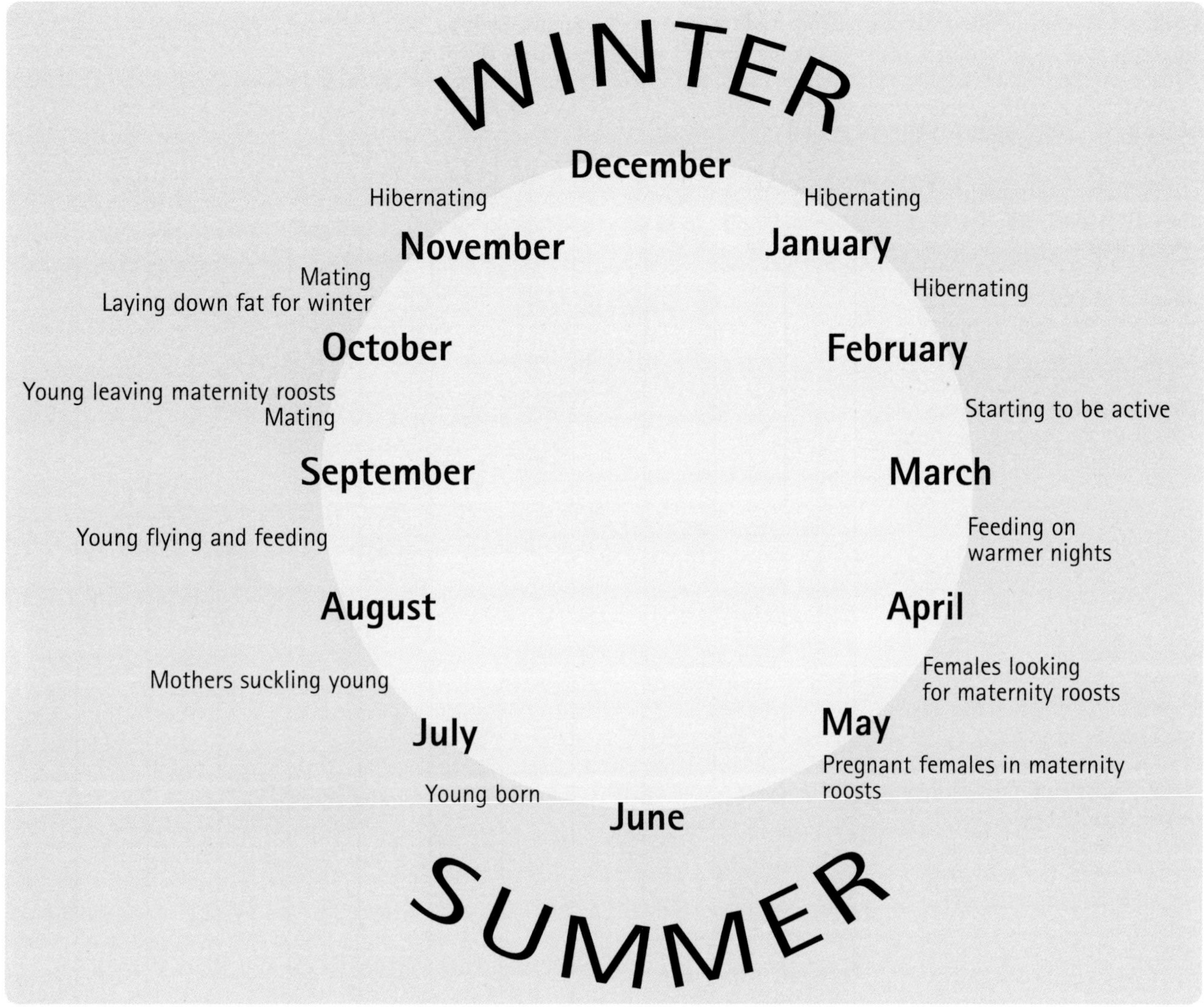

The timing of surveys depends in part on the ecology and behaviour of bats, which changes markedly throughout the year and influences where and when they may be found in different habitats. See Figure 3.2 for a brief overview of a bats' annual cycle.

A large part of the bat year, October to May, can be spent in hibernation although on warm nights bats awaken to go out and forage.

As spring approaches, bats increasingly go out to feed and the period from May to early June is a time of intense feeding activity to recover weight lost during the winter. During this time, females gather together at maternity roosts, sometimes moving from one roost to another to find one with the right conditions. Some males may be present, but most roost elsewhere, either singly or in small groups.

Most births occur in the middle two weeks of June but, depending on the weather, this can be as early as late May or as late as August. Once the babies are weaned, at three to five weeks, the females leave the maternity roost and disperse, both to gain weight before winter and to find mates.

During autumn, many *Myotis* bats travel to underground sites for an activity known as swarming which may be concerned with mating and/or finding a hibernaculum. Males of other bat species establish mating territories where they may fly or call specifically to attract a mate.

As the weather turns colder, bat activity reduces and foraging becomes restricted to warmer nights. Bats spend progressively more time in torpor, before returning to hibernation.

Table 3.3 provides recommended optimal and sub-optimal timings for all types of survey methodology discussed in these guidelines. Although the survey timings are applicable for most of the UK, they will vary according to the geographic location of the survey, with bat activity in Scotland commencing later than in most of England. They may also vary from year to year depending on the timing of the onset of spring. As outdoor surveys are weather dependent, all surveys should be undertaken in suitable weather conditions.

Table 3.3 Recommended survey periods (adapted from Limpens, 2005)
Site walkovers and scoping surveys can be carried out throughout the year.

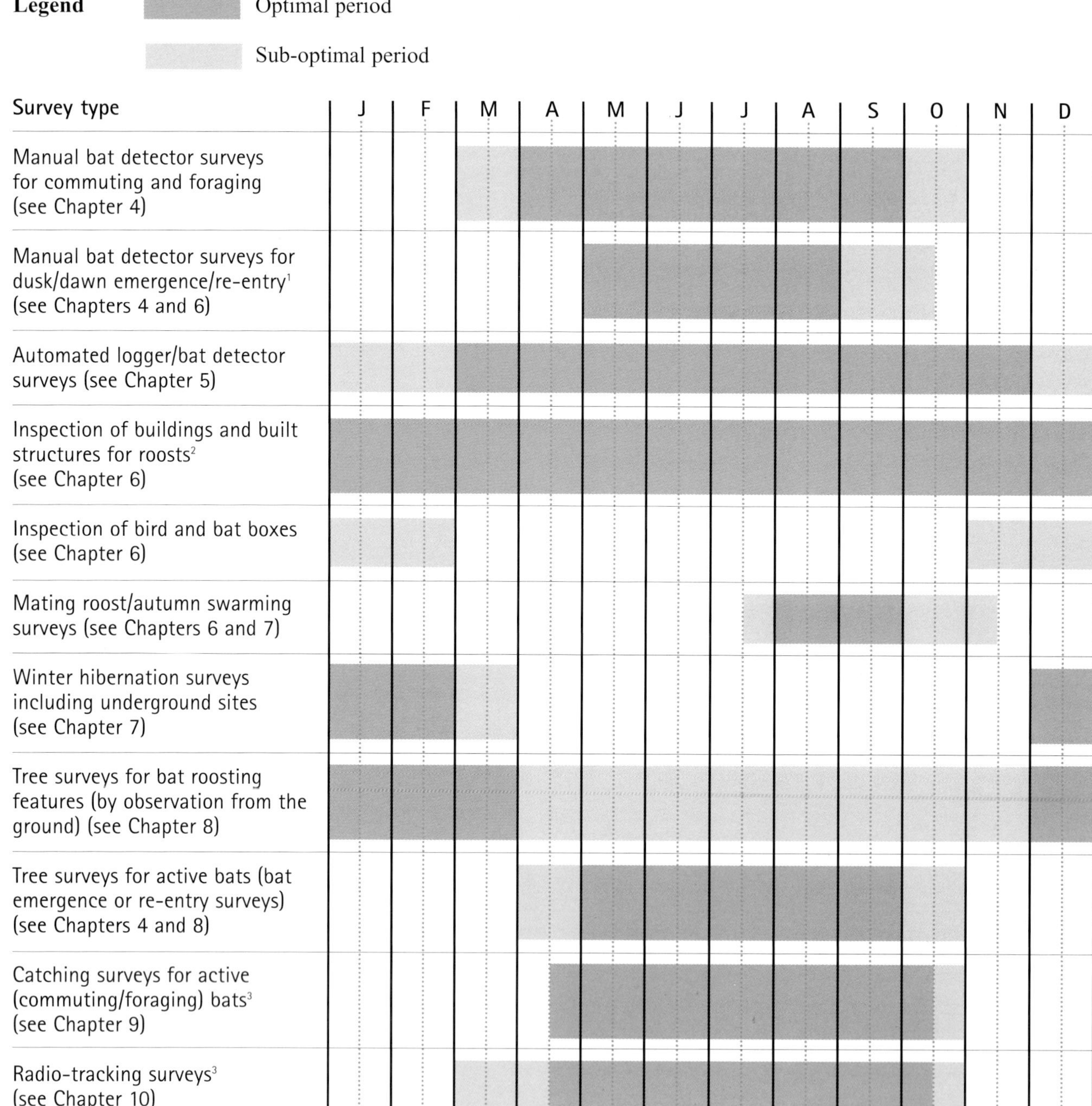

Notes

1 The months shown are optimal for maternity roosts but there are other types of summer roosts for which other months outside these may also be optimal.

2 Although inspections of buildings can be made year round, summer roosts are more easily found during the months of May to August.

3 As best practice guidance the period for catching and radio-tracking bats should avoid periods of late pregnancy/early lactation or early flying juveniles, except in exceptional circumstances when the survey specifically requires it.

3.10 Recording and reporting surveys

It is essential during planning and preparation of surveys to give due consideration to the recording and reporting of survey data. The results must be recorded in a thorough and clear manner and it is recommended that a standard format is used. A general reporting template is given in Box 3.2. Not all parts will be relevant in every case so it should be modified and used accordingly.

All survey records should include the following:

- date of the survey;
- location of the survey (including six-figure grid references);
- list of the personnel who undertook the survey, their qualifications and experience, including details of relevant licences; and
- species identified (supported where possible with objective evidence e.g. sonograms).

In addition, reports of field surveys should include:

- weather conditions measured at intervals through the survey (temperature, humidity, wind speed and direction, cloud cover, precipitation and light levels); and
- annotated diagrams and/or tables of bat activity and transect routes showing species encountered, time, location and activity (e.g. direction of flight, feeding, commuting or other activities).

Further examples of the records appropriate to different types of survey are given in later chapters. A reporting template for roost visits can also be found in Appendix 3 of the *Bat Workers' Manual* (Mitchell-Jones and McLeish, 2004).

Survey reports must make clear the limitations and constraints of survey methodologies employed and explain the impact of these on the conclusions that can be drawn from the survey data. Frequency of visits and survey staff might change for a variety of reasons and when this occurs it is important to report it and interpret the findings accordingly. It is also important to report and justify a reduced (or increased) survey effort, if that was the case.

It is good practice for bat survey records resulting to be submitted to Local Records Centres, local bat groups and/or the National Biodiversity Network (**www.searchnbn.net**).

Once the first draft of a report has been written, comments should be sought (see Section 3.11).

Box 3.2 Reporting template
This is a general reporting template; not all parts will be relevant in every case so it should be modified and used accordingly.

1 INTRODUCTION

1.1 **Site description**
Six-figure grid reference, description of location, brief habitat descriptions (including presence of buildings or other features of potential value to bats), size of site and photo/plan of site.

1.2 **Proposed works**
If a development, what is proposed in terms of demolition, habitat destruction, renovation, timings, etc. If details of proposed works are not known, then state this here.

1.3 **Aims of study**
Objective of the survey undertaken and the aims of the report.

2 METHODOLOGY

2.1 **Desk study**
List the organisations and sources from which bat records and information on sites of nature conservation importance have been requested and obtained. If no data sought or secured, then state this.

2.2 **Field survey**
Names of surveyors, their NE/CCW/SNH/DoENI bat licence numbers or other evidence of competency if surveyor is not licensed.
The date of survey, time of start and end, weather conditions (temperature, humidity, wind speed and direction, cloud cover, precipitation and light levels).
Survey methodology used, with reference to standard published methodologies, and way in which this seeks to achieve the survey objectives.
Make clear which parts of a site were surveyed (e.g. survey of buildings only) and record the equipment used by each surveyor.

3 RESULTS

3.1 Desk study

Results of data trawl if sought and obtained.

3.1.1 Designated sites

Location of sites designated for their bat interest.

3.1.2 Protected species

Location and age of bat records.

3.2 Field survey

This information may be presented in many forms; as appendices, tables or text.

3.2.1 Habitat description

Description of habitats of value to commuting, foraging and roosting bats.

3.2.2 Bat roost survey

Full detailed descriptions of features that have potential value to bats, including location of potential access and roosting points as well as actual access and roosting places and descriptions of bat signs and live bats found. This section may not be relevant if activity surveys only were undertaken.

Location of roosting features, height above the ground, direction in which they face and description of features with potential to support bats as well as actual bat roosts.

Building/built structures - type of building/built structure, dimensions, age, construction, current usage.

Underground site – type of site (e.g. cave/mine and dimensions).

Trees – species, dimensions, age, details of roost access.

3.2.3 Bat activity survey

Results of transect surveys, including analysis of bat detector recordings as well as sightings during survey.

Results of emergence/re-entry surveys (if appropriate).

Results of automated bat detector/logger surveys (if appropriate).

Results should all be both quantitative and qualitative.

4 ASSESSMENT

4.1 Constraints on study information

Weather, access, health and safety or any other constraints which have resulted in an inability to undertake a full survey and or make a full assessment of impacts.

4.2 Potential impacts

Impacts considered both during construction and in scheme operation.

4.2.1 Designated sites.

4.2.2 Bat roosts.

4.2.3 Bat foraging and commuting habitat.

4.3 Legislation and policy guidance

Details of UK and European legislation relating to bats, relevant national and local planning policy, national and local bat species biodiversity action plans.

The purpose of this section is to put the findings of the survey and the impact assessment into a legal and policy context.

5 RECOMMENDATIONS AND MITIGATION

5.1 Further survey

5.2 Mitigation measures (not applicable to scientific/conservation surveys)

5.2.1 Mitigation for roost sites.

5.2.2 Mitigation for foraging and commuting habitat.

5.3 Requirement for Habitats Regulations (EPS) licences

If bats or their places of shelter or protection are likely to be affected by the proposed development then consideration may need to be given as to whether a Habitats Regulations licence should be sought.

6 SUMMARY

A brief precis of the report's findings.

7 REFERENCES

APPENDIX 1. Desk study data

APPENDIX 2. Bat roost report forms

APPENDIX 3. Photographs

3.11 Consultees

Consultations may be required at various stages of a research, conservation or development project. These stages may be:

1. during development of survey methodology;
2. following survey findings;
3. prior to, or following, submission of a planning application; and/or
4. when planning conditions are to be fulfilled.

Consultees may include any of the following:

- Countryside Council for Wales;
- Environment Agency;
- Environment Agency Wales;
- Environment and Heritage Service Northern Ireland;
- local and national non-statutory nature conservation bodies such as local bat groups and wildlife trusts and other relevant associations, for example, the British Caving Association in cases involving underground sites;
- Local Planning Authorities;
- Natural England;
- Scottish Environment Protection Agency; and
- Scottish Natural Heritage.

The form of consultation will vary with the purpose and type of survey. In all cases, SNCOs must be informed about any proposals. Where certain species or protected habitats are involved, a more detailed discussion with the SNCO may be useful or required. In most cases it will be for the surveyor to provide the detail of, and justification for, the survey, but valuable suggestions for improvements may also be obtained.

Chapter 4

Manual bat activity surveys

4.1 Introduction

In this chapter, manual bat activity surveys are defined as surveys undertaken by people, the surveyors, observing bats outdoors in the field. The surveys do not disturb bats and are therefore non-invasive.

The surveyors usually work as a team, focusing on observing and recording bat activity in a given area. Surveyors should always carry a bat detector, which greatly increases the ability to discover bat presence and can assist in identifying the species. Night vision equipment can also help, allowing observations of bat activity in the dark.

Surveys using equipment such as automated loggers or detectors left in the field are discussed in Chapter 5. In practice, non-invasive surveys of bat activity may use a combination of more than one method (manual and automated) to maximise efficiency.

For information on assessing the need for a bat survey and its planning and preparation refer to Chapters 2 and 3.

4.1.1 Understanding bats' use of the landscape

When planning and undertaking manual bat activity surveys it helps to know when and how bats use the landscape, and this requires an understanding of bat ecology (see Figure 3.2 in Chapter 3 for a brief overview of a bat's annual cycle). Bats are highly mobile animals travelling between:

- hibernacula and summer roosts;
- one summer roost and another;
- roosts and feeding areas; and
- roosts and mating sites.

Surveys should always be set in the regional and seasonal context; for example, maternity roosts are not found in the UK in October. Understanding bat behaviour is also vital to interpreting correctly the findings of the survey at a later date.

British bats show a preference for habitats associated with broadleaved woodland and water. In addition, they select linear features in the landscape (hedgerows, tree lines, river and streams), not only for commuting but also for foraging. This is particularly true of the *Myotis* species and pipistrelles. Bat activity is dictated by the location and emergence of insects; bats know when and where is their best foraging. This explains why seemingly good bat habitat sometimes has no bats, because they are feeding elsewhere.

Taking account of bat ecology, Table 4.1 gives information, where possible, for individual species in the UK in terms of foraging and commuting habitat, emergence time, range and distribution.

Table 4.1 provides a broad description of bat behaviour but should only be used as a guide. Professional surveyors should always keep abreast of current knowledge and research. Moreover, as with most ecological survey work, there is no substitute for practical experience. With greater experience comes an improved sense of where bats are most likely to be within a landscape. However, experience can also lead to over-confidence, which can result in some potential habitat being dismissed without a survey.

This chapter aims to provide guidance on the approach and survey effort required to assess bat activity. The guidance should be adaptable to most site survey situations where an assessment of bat activity in the field is required.

Table 4.1 Foraging and commuting habitat, emergence time, flying range, and distribution of resident breeding bat species in the UK[18]

Species	Foraging and commuting habitats, emergence time, range and distribution
Common pipistrelle *(Pipistrellus pipistrellus)*	**Foraging and commuting habitat:** Exploit a wide range of foraging habitats including those associated with watercourses, woodland, grassland and built-up areas; may also feed around lighting. **Emergence[1]:** From approximately 30 minutes after sunset. **Flying range:** Feeding areas up to 3 to 4 km from their roost. **Distribution:** Common throughout the UK and is the most commonly found bat in England and Wales.
Soprano pipistrelle *(Pipistrellus pygmaeus)*	**Foraging and commuting habitat:** Forage mostly over habitat associated with water and follow water features when they travel. **Emergence[1]:** From approximately 30 minutes after sunset. **Flying range:** Feeding areas up to 3 to 4 km from their roost. **Distribution:** Common throughout the UK. It is the second most commonly found species but is more abundant in the north and west.
Nathusius' pipistrelle *(Pipistrellus nathusii)*	**Foraging and commuting habitat:** Forages over water and also along woodland edges and rides. **Emergence[1]:** From approximately 30 minutes after sunset. **Flying range:** Nightly flying range to feeding areas is poorly known. European data demonstrate this bat is a long distance migrant (> 1,900 km recorded) and may cross from continental Europe to the United Kingdom seasonally. **Distribution:** Information is currently too limited to give a distribution.
Brown long-eared bat *(Plecotus auritus)*	**Foraging and commuting habitat:** Live and forage around woodland and parkland with old trees. **Emergence[1]:** When dark, from approximately 60 minutes after sunset. **Flying range:** Generally feed within 1-2 km of the roost; some bats will travel further. **Distribution:** Common throughout UK wherever there is suitable woodland.
Grey long-eared bat *(Plecotus austriacus)*	**Foraging and commuting habitat:** Forage in more open areas (when compared to the brown long-eared), over grassland and along woodland edges. **Emergence[1]:** When dark from approximately 60 minutes after sunset. **Flying range:** Feed up to 6 km from the roost. **Distribution:** A rare bat found in a few areas of southern England and in the Isle of Wight.

18 Key references: Altringham (2003), Boye and Dietz (2005), Hutterer *et al.* (2005), Russ (1999) and Schofield and Mitchell-Jones (2003).

Natterer's bat *(Myotis nattereri)*	**Foraging and commuting habitat:** Hunts in tree canopies or close to foliage and by edges of water although at a higher level than Daubenton's bat. **Emergence[1]:** From approximately 40 to 70 minutes after sunset. **Flying range:** Generally feed up to 3 km from the roost. Recorded to travel around 60 km between summer and autumn/winter sites. **Distribution:** Found throughout UK with the exception of the far north of Scotland; wherever there is suitable woodland.
Whiskered/Brandt's bats *(Myotis mystacinus/Myotis brandtii)*	**Foraging and commuting habitat:** Whiskered bats forage in a wide range of habitats including parkland, woodlands, flowing water and suburban gardens. Brandt's bat forages more in woodlands and close to water bodies. **Emergence[1]:** From approximately 30 minutes after sunset. **Flying range:** Flying range is poorly understood. The distance between summer and winter roosts is usually less than 50 km. **Distribution:** Little is known about their individual distributions. Both species found throughout England, Wales and southern Scotland and parts of Northern Ireland.
Daubenton's bat *(Myotis daubentonii)*	**Foraging and commuting habitat:** Hunts close to the surface of slow-moving or calm water. Will also forage in trees or along woodland rides, especially if these are associated with water. **Emergence[1]:** From approximately 40-60 minutes after sunset. **Flying range:** Mostly feeds up to 6-10 km from the roost. Have been recorded to travel much further (100 km) between summer and winter roosts. **Distribution:** Found throughout the UK, with the exception of some offshore islands.
Bechstein's bat *(Myotis bechsteinii)*	**Foraging and commuting habitat:** Forages in areas of closed-canopy woodland close to water. It will also forage along overgrown hedgerows and tree lines. **Emergence[1]:** Early evening around sunset to 60 minutes after sunset. **Flying range:** Typically feeds within 1 km of the roost. Probably travels further between summer and winter roosts. **Distribution:** Southern England from Sussex to Gloucestershire; has been recorded in Shropshire and occasionally in Wales. Not recorded in Scotland or Ireland.
Noctule *(Nyctalus noctula)*	**Foraging and commuting habitat:** Flies high and straight to feeding sites over parkland, pasture, water and deciduous woodland. May feed around lighting. **Emergence[1]:** Early evening in daylight; approximately sunset. **Flying range:** European data demonstrate this bat is migratory (> 1,000 km recorded). **Distribution:** Found throughout England and Wales and into southern Scotland. Not recorded in Ireland.

Leisler's bat *(Nyctalus leisleri)*	**Foraging and commuting habitat:** Flies high and straight. It forages mostly in areas of open habitat often over water or pasture. **Emergence[1]:** Early evening from approximately 15 minutes after sunset. **Flying range:** Feeds up to 14 km from its roost. European data demonstrate this bat is a long-distance migrant with regular seasonal movements (>1,000 km recorded). **Distribution:** Common and widespread in Ireland, rare and widespread in England and southern Scotland with a few records in Wales.
Serotine *(Eptesicus serotinus)*	**Foraging and commuting habitat:** Forages mainly over pasture, parkland and along woodland edge. May also feed in suburban gardens and around street lamps. **Emergence[1]:** Early evening from approximately 15 minutes after sunset. **Flying range:** Typically forages within 2–6 km of roosts. **Distribution:** Widespread but scarce in southern Britain; restricted to the south and south-west of England with occasional records in Wales. Not recorded in Scotland or Ireland.
Barbastelle *(Barbastella barbastellus)*	**Foraging and commuting habitat:** Forages along woodland edge and rides. Uses hedgerows, streams and rivers for commuting. **Emergence[1]:** From 20-60 minutes after sunset. **Flying range:** Typically forages 5-6 km from their roosts; has been recorded travel up to 18 km. **Distribution:** Widespread. Recorded from a line south of North Wales to the Wash. Not recorded in Scotland or Ireland.
Greater horseshoe bat *(Rhinolophus ferrumequinum)*	**Foraging and commuting habitat:** In spring, greater horseshoes feed over cattle-grazed pasture and in ancient or semi-natural woodland. During summer, they forage over hay meadows and silage fields. **Emergence[1]:** From approximately 25-50 minutes after sunset. **Flying range:** Typically feeds within 4 km of roosts but when habitat is restricted travel in excess of 12 km. Occasionally moves more than 100 km. **Distribution:** Found in south-west England and south and west Wales, with an increasing number of occasional records in mid- and North Wales. Not recorded in Scotland or Ireland.
Lesser horseshoe bat *(Rhinolophus hipposideros)*	**Foraging and commuting habitat:** Flies to feed over areas of deciduous woodland, especially closed canopy. Will feed during the winter. **Emergence[1]:** From approximately 30-50 minutes after sunset. **Flying range:** Typically feeds within 2 km of their roost. Will travel 20 km between winter and summer roosts. **Distribution:** South-west England, Wales and western Ireland. Not recorded in Scotland.

Note

1 Time of emergence can vary through the year; it is also affected by the weather and the cover of local habitat.

4.1.2 Aims of survey

The aims of manual field surveys are to:

- determine presence/absence of species i.e. the species present in a given area;
- determine the intensity of bat activity both spatially and temporally (to help estimate bat populations);
- determine the type of activity, most usually:
  - foraging (by feeding buzzes);
  - commuting (by high directional pass rates);
  - mating (by mating social calls);
- find roosts by tracking back bat flight paths or observing dawn flight activity at roosts; and/or
- find emergence of bats from a building or built structure.

It should be noted that absence is near impossible to prove, as there is always a chance of finding a bat. Their high mobility means that it is virtually impossible to rule out bats using any type of structure for roosting or habitat for foraging or as a flight path. When there is no conclusive evidence of bats, a potential for bat presence should be given.

These surveys are undertaken outdoors and the surveyor must have considerable field skills to be competent in finding and identifying bats and interpreting their activity. A licence is not required to undertake these surveys and it is worth noting that the training required for a roost visitor or survey licence does not usually provide detailed tuition on the use of bat detectors and species identification necessary for manual field surveys. For skills required and details of training available see Sections 3.2 and 3.3 respectively.

The information that can be obtained from non-invasive detector surveys (manual or automated) should preferably be exhausted before any invasive techniques, such as mist netting or radio-tracking, are used (see Section 3.8.1).

Activity surveys may be required for a range of habitats and for varied development proposals, both of which may influence the way in which a survey is conducted and the survey methods used. It is important to note that all manual bat activity surveys are just a snapshot of bat activity in a small space at a particular time. To help overcome these spatial and temporal constraints and to maximise the survey effort, the following approaches are used:

- surveys are conducted when the bats are most likely to be active (during warmer months and at dusk and dawn); and
- surveys concentrate on habitats and features that are likely to be used by bats.

This is applied in practice by:

- undertaking surveys at dusk and at dawn;
- spreading the individual surveys throughout the summer months; and
- performing transects over the area of interest or being stationary at specific sites while observing and recording bat activity, or a combination of continuous/line transects and point counts. Surveying by car or boat can also increase the area covered compared with surveying on foot (see Sections 4.6.2 and 4.6.3).

To date, there is no statistically robust survey method for determining the bat species, population and level of activity in a given area. Moreover, consideration must be given to determining bat activity on a large scale and over an extended period when the potential effects of the development are not fully understood or have not previously been assessed, for example, with wind farm proposals.

4.1.3 Identifying bats by echolocation and flight

Identification of bats in the field is aided by careful recording (e.g. a sound recording with commentary) of as many factors as possible including:

- echolocation calls;
- pattern of flight;
- apparent size;
- location and habitat;
- type of activity – feeding, commuting, social calling;
- time of activity; and
- weather conditions.

The value of field experience cannot be over-emphasised. Even with moderate experience it is possible to use bat detectors to separate bats into four groups based on the type of echolocation call they use (particularly the frequency ranges and relative amplitudes involved) although the information gained will be of limited value. The four groups are:

- horseshoe bats;
- lower-frequency bats (also sometimes referred to as the bigger bats): noctule, serotine and Leisler's bat;
- pipistrelle bats; and
- everything else (which includes barbastelle, long-eareds and *Myotis* bats).

With further experience these groupings can be split further using heterodyne or tuneable detectors to listen to the spacing between calls, the relative loudness and the change in pitch or note of the call as you tune through its frequency range. These call parameters can also be measured if you record the calls with a broadband bat detector (frequency division or time expansion) and analyse them using appropriate software. As well as sound, the manner, height and location of flight along with the habitat and time of observation will assist with identification. Be aware that results from detector encounters are biased towards bats that use louder and lower frequency calls. Quiet bats like Bechstein's may be present and not picked up by a detector unless they are a few feet away. Appendix 1 contains more detail on how to identify bats using ultrasound calls (including sound analysis) and flight patterns.

When using a bat detector to survey, well-charged batteries are essential as low batteries severely affect the sensitivity of a bat detector. The geographical location should be considered and notes about the surroundings and flight pattern should be taken, if post-survey sound analysis is to be carried out. Ideally, the analysis should be done as soon as possible after the survey. Sound analysis is a useful tool and may help to identify species otherwise missed, but it has its limitations, especially for bats in cluttered habitats and where species have overlapping call parameters such as *Myotis* species.

It should be noted that social calls are more common in autumn than at other times of year. Pipistrelle social calls sound a lot like noctule calls to the inexperienced surveyor using a heterodyne detector. Pipistrelle social calls also tend to travel further because they are low frequency, so they may be heard without an associated typical echolocation call. Mothers may make distinctive sounds towards their young, and juveniles may also produce atypical calls when learning to echolocate. For identifying social calls, the use of time expansion detectors to make recordings for sound analysis is advised.

When surveying for bats always be prepared for the unexpected; bats do not read the guidance and there is still a lot to learn.

4.2 Equipment

Appropriate equipment is essential for surveying bats; the bat detector is the most useful tool and no manual field survey for bat activity should be attempted without one. Bat detectors use three methods to convert bat ultrasound to a range audible to humans; the advantages and disadvantages of each type of detector are given in Table 4.2, along with those of night vision scopes and low-light video cameras.

It is good practice to make a sound recording of the survey using a broadband detector for future analysis. Analysis of the broadband recording on a computer enables objective evidence to be produced to support field observations.

Equipment to measure key weather parameters, especially at the microclimate level, is also needed. Essential and recommended equipment is listed in Table 4.3 and Table 4.4 respectively.

Equipment often fails, so spares should naturally form part of the survey kit. For manual field surveys always carry a spare bat detector, torch and batteries.

Manual bat activity surveys are limited by the equipment used. Detection methods bias the results of observations in many ways:

- Bats vary in the loudness of their call so louder bats are more likely to be detected by a bat detector.
- The area and range of detection equipment such as night vision and bat detectors is limited, so bats high in the canopy are less frequently recorded.
- Microphones in bat detectors vary in sensitivity so if a bat detector is changed for a more sensitive model, this can cause an apparent increase in the number of bats recorded.

Potential bias caused by manual bat detection methods should be taken into account when interpreting the results. If the bias cannot be adjusted (or qualified) through expert judgement and there is a need to confirm the presence of individual species, rather than infer their presence by the habitat or nearby records, or the mitigation cannot be undertaken without confirming their presence, then the manual bat activity survey techniques should be supplemented by other survey methods (such as capture and possibly also radio-tracking - see Chapters 9 and 10) subject to the constraints detailed in Section 3.8.

Table 4.2 Advantages and disadvantages of different detection methods for manual bat activity surveys including bat detectors, night vision scopes and video cameras.

Detector type	Advantages	Disadvantages	When best used
Heterodyne detector	• Most familiar type of bat detectors and pleasant to listen to. • With expertise can identify most species in the field at least to species group.	• No sound analysis possible. • Narrow band listening (species outside listening range can be missed). • Louder bats bias the estimate of population as they are the ones most often heard.	• Non-professional surveys or in professional surveys when in combination with broadband detector types.
Frequency division detector (broadband)	• Broadband recording (all species heard at once) with some frequency division detectors (e.g. Petersson D230 and Batbox Duet). • Records continually through the survey so nothing is missed. • With sound analysis, can be used for identifying some bats to species level and others to species group. • Many more calls recorded in a given time period than with time	• Hear all the ultrasound clutter. • Sound analysis not as good as that produced from time expansion using spectrographic methods as the sonograms contain less information. • Louder bats bias the estimate of population as they are the ones most often heard. • Low frequency, frequency-divided calls contain little information for the sonogram.	• Professional consultancy surveys in conjunction with recording equipment when identification to generic level is adequate, especially when quantity rather than quality is important. • In combination with heterodyne or time expansion detector.

	expansion and hence less likelihood of identification based on anomalous calls. Preservation of repetition rates and rhythms to aid recognition in the field by some detectors (e.g. Petersson D230 and Batbox Duet).	• Only one harmonic is sampled. • Cannot replay the sound in heterodyne mode (although some sound analysis software will re-create the sound).	
Time Expansion detector (broadband)	• Broadband recording (all species heard at once). • Preserves the detail of the high frequency sound. • With sound analysis can be used for identifying some bats to species level. • Can replay the sound in heterodyne mode (e.g. Petersson D240x) or through some sound analysis software.	• Does not continually record, so some bats may be missed. However, if the recording time is kept short (<100 ms), the detector will catch a selection of calls from most passes though there will be a trade off in terms of identification potential. • False triggers by non-bat ultrasound in automatic recording modes. • Resolution not sufficient for differentiating whiskered, Brandt's and Bechstein's and can be limited in certain situations for other species. • Louder bats bias the estimate of population as they are the ones most often heard.	• Professional research work when identification to species is necessary and the details of the call are essential, especially where quality rather than quantity is required. • Professional consultancy surveys in combination with other detectors when selected calls may be recorded for analysis.
Night vision scope	• Can observe bats that cannot easily be heard. • Can observe bat flight behaviour and its location.	• Can have a narrow field of view making it difficult to follow bats in flight. • Cannot pan quickly.	• Used to observe bat flight behaviour (e.g. on exit surveys when the exit is dark). • Best used in conjunction with an infrared light source.
Video camera (with low light recording capability)	• Provides a visual record. • Can observe bats that cannot easily be heard. • Can observe bat flight behaviour and its location.	• Cannot pan quickly.	• Used to observe bat flight behaviour (e.g. on exit surveys when the exit is dark). • Can act as additional observer. • Best used in conjunction with an infrared light source.

Table 4.3 Essential equipment for manual bat activity surveys

Equipment	Use
Bat detector	Assists in finding bats.
Recording equipment (various types - analogue and digital)	Sound record of survey for subsequent computer analysis.
Headphones	Greatly increases the sound quality on the detector - also allows two methods of listening to bats (e.g. heterodyne and frequency division).
Computer with spectrographic sound software	Spectrographic sound analysis.
Hand-held temperature and humidity meter	Instruments to record temperature and humidity (wind speed and direction and cloud cover can be estimated by eye).
Plans	To record findings.
Torch/head torch	To light the way, for safety and to aid recording notes.
Compass	For orientation.

Table 4.4 Other equipment useful for manual bat activity surveys

Equipment	Use
Clinometer	Measuring the height of a roost.
Night vision equipment	Viewing bats in the dark (e.g. to observe foraging in the canopy).
Two-way radio (walkie-talkie)	Communication with other surveyors and to hear progress of survey. Channel can remain open for continuous listening.
Global Positioning System (GPS) or Satellite Navigation System (Sat Nav)	Accurate positioning and recording of time and location on driven transects.
Lux meter	Measure ambient light levels.
Torch with infrared filter	Used in conjunction with night vision allows for non-invasive viewing of bats.
Tally counter	For emergence counts.
Close-focusing binoculars	Inspection of potential bat features.
Voice recorder	Making notes of survey (some bat detectors have this feature).
Watch	Stopwatch and recording times.

4.3 Timing

Manual bat activity surveys should be programmed to take place when bats are most likely to be active. Start times will vary throughout the year, according to the dusk and dawn times.

In practice, surveys are normally undertaken as:

- dusk survey only;
- dusk and dawn surveys with a rest break between the two;
- dusk to dawn surveys (as undertaken for radio-tracking studies but can provide valuable information when conducted as manual activity surveys); or
- dawn survey only.

Table 4.5 gives guidance on the period of time the survey should be undertaken for a series of common survey objectives.

Table 4.5 Recommended length of time over which manual bat surveys should be conducted

Survey objective	Dusk survey[1]	Dawn survey (if undertaken)
Bat activity away from roost (all species)	sunset to 2-3 hours after sunset[2]	2 -1 1/2 hours before sunrise to sunrise[4]
Bat emergence from and re-entry to roost (all species)	1/4 hour before sunset to approximately 2 hours after sunset[3]	2 -1 1/2 hours before sunrise to sunrise[4]
Mating activity (all species)	sunset to 4 hours after sunset	-

Notes

1 These surveys are not normally undertaken as one-offs and are usually part of a series see Table 4.7.

2 When the site area larger than 1 ha and within 4 km of a greater horseshoe bat roost, 3 hours is required.

3 Some bats may emerge earlier and 1/2 an hour before sunset may be more appropriate.

4 Some bats may return to their roost after sunrise.

4.4 Survey area

The area of coverage for the survey depends on the following:

- size and complexity of the development;
- proximity of designated sites (e.g. SAC or SSSI);
- value of surrounding habitats for bats; and
- known bat populations in the area.

As a starting point, it is recommended that when assessing high or moderate value sites (see Box 3.1) the survey should extend to include all structures and areas that are affected by the development and which are assessed (after the data trawl and walkover - see Sections 3.6 and 3.7) as having the potential to be important for bats. The survey may need to extend beyond the site boundary; for example, light spillage may disturb a flight corridor outside the development. The survey may also need to be extended when there is the potential to affect designated sites or other known bat populations. For phased developments, the entire site should be surveyed at the start.

4.5 Survey effort and frequency

Having defined the survey area, deciding on the number of surveyors per unit area is more difficult. Table 4.6 gives guidance on the number of surveyors per development area (ha). However, the optimum number of surveyors will depend not only on the area of development but also on the bat potential of the site in terms of the species expected to be present, the number of potential roost sites and the availability of suitable habitat. Ease of access and navigation around the site may also be a factor in determining the number of surveyors; for example, a site with many buildings will require more surveyors to survey for emergence activity. The number of surveyors will also be influenced by the type and number of bat detectors available plus the objective of the survey. More heterodyne detectors will be needed to cover an area thoroughly than if using broadband detectors. The bat potential, and hence the required number of surveyors, is decided after the data trawl and preliminary walkover survey.

When undertaking surveys for development, the level of survey effort should be proportionate to the likely use of the site by bats and the potential effects of the proposed development on the species present. The effort is closely related to the habitats present; for example, more effort is required when the site is predominately deciduous woodland and a river corridor than if it were mostly a paved car park. However, the regional location of the site also needs to be taken into consideration, with increasing species diversity (and perhaps also density) to the south and west of the UK.

Table 4.6 Recommended number of surveyors per site area using broadband detectors (Not applicable to emergence surveys)

Site area (ha) *The site has a moderate likelihood of bats (see Box 3.1)*	**Number of surveyors**
< 5	2
5 to 25	4
25 to 75	6
75 to 200	8

Notes on the number of surveyors

1 Two surveyors is the recommended minimum for health and safety reasons; surveyors may have to work in pairs to satisfy the risk assessment.

2 If the required number of surveyors is not available, then the survey should extend over several nights to keep the area /surveyor ratio roughly constant.

3 Automated systems (e.g. data loggers) can be used in place of some of the surveyors. This is especially recommended for large areas as this can be an easy way to extend the survey period.

4 Driven transects use fewer surveyors per unit area; however, the value of these is not equal to that of a walked transect, which would be undertaken over the same length of time but in a smaller area.

In practice, manual bat activity surveys are undertaken on sites that are generally a mixture of habitats and the survey effort is decided using the following factors:

- area and value of the habitat which is of potential use by bats (hedgerows, tree lines, river or streams, etc.);
- location (to the south and west of the UK there are more bat species and perhaps a higher number of bats per unit area); and
- linkage to potentially good bat habitats outside the development area; these may be many kilometres away. A thin corridor may be important to commuting bats.

Table 4.7 gives guidance for the minimum number of manual bat activity surveys that should be conducted in order to ensure that sufficient survey visits are undertaken to observe bat activity and to have confidence in a negative result. In this table, site habitat has been simplified into moderate to high and low value. This should have been indicated by the data trawl and the walkover. These should also have indicated the potential presence of bat roosts.

Note that the proximity of a greater horseshoe bat roost can override the site value of the habitat, and there is separate published guidance for greater horseshoe bat surveys in the *Bat Mitigation Guidelines* (Mitchell-Jones, 2004).

Table 4.7 Minimum visit frequency and timing for manual bat activity surveys away from and at roosts (see also Table 3.3 for recommended survey periods and Box 3.1 for guidance on habitat values)

Location of activity survey	Site/habitat being surveyed	
	Moderate to high value	Low value
In habitat away from known roosts (bat detector transects) (all species)[4]	2/3 surveys[1] during March[2] - September[3]. Optimum period June - August. At least one of the three surveys should comprise dusk and dawn (or dusk to dawn) within one 24-hour period.	2/3 surveys[1] during March[2] - September[3]. Optimum period June - August.
At known roosts (dusk emergence or dawn-re-entry surveys) (all species)	2/3 surveys[1] during May - September. Optimum period May - August. At least one of the three surveys should comprise dusk and dawn (or dusk to dawn) within one 24-hour period.	2/3 surveys[1] during May - September. Optimum period May - August.

Notes

1 Best practice is to space the surveys evenly through the optimum period.

2 Surveys should not start before April in the north of the UK; some delay may also be caused by the weather.

3 Season can be extended to October to November if working in south-west England or surveys for mating activity are required, although particularly cold weather will render this inadvisable.

4 When the site area is larger than 1 ha and within 4 km of a greater horseshoe bat roost, the *Bat Mitigation Guidelines* (Mitchell-Jones, 2004) recommend two surveys in each month between May and September. Advice from the SNCOs should be sought and the same may be applied to other species also (in particular, other Annex II Habitats Directive species).

4.6 Methods

Various methods can be employed during manual bat activity surveys in order to gain an understanding of how bats use an area. A single survey method is rarely used in isolation; for example, walked and driven transects may both be carried out at the same time and complemented by automated survey techniques (see Chapter 5). It is recommended that driven transects are used to supplement rather than replace walked transects, as few proposed developments are able to be wholly surveyed from a road. Moreover, the road network must be such that it would provide useful information about bats in the area.

Peak activity is seen at dusk and dawn and there can be a lull in activity in the early hours of the morning, although at autumn swarming sites bats remain active for several hours after dusk (see Chapter 7). For this reason surveys are usually concentrated in the hours after dusk. However, it is recommended that, as a minimum, one dawn survey is undertaken as this invariably increases the understanding of how bats use the site and can be particularly useful in locating roosts by detecting bats engaging in dawn swarming activity outside the entrances to their roosting sites.

4.6.1 Walked transects

Preparation

The chosen route should not be walked for the first time in the dark. The site should first be visited in daylight to plan the route and 'listening station' stops, and to walk the whole route, noting how long one circuit takes. The aim is to ensure that the whole transect takes no more than three hours starting about half an hour before sunset. If the site to be surveyed is small, more than one circuit may be achieved in one survey session. At a large site, transects may need to be undertaken over several consecutive nights to cover the whole area.

Using aerial photographs, an Ordnance Survey map and a daytime walkover, a route should be chosen that will incorporate habitat features with potential for use by foraging and commuting bats. These may include, if present on the site, woodland, woodland edge, hedgerows, lines of trees, stream corridors, lake or pond edges, scrub margins and grassland, especially semi- or unimproved pasture.

Method

Transects are walked usually with a broadband bat detector connected to a recording device, or the recording device may be an integral part of the detector. Heterodyne detectors cannot be used for analysis of frequency information. Some detectors, such as the automated Anabat or the Tranquility, can be hand-held throughout the walked transect, leaving them to record automatically or when ultrasound triggers the recording. Recordings can include the exact time of the bat pass and these can be checked against the surveyor notes during later analysis. Other detectors can be set to record either automatically at an ultrasound trigger, or used on a manual setting. In manual mode, the observer pushes a button to start the sample when ultrasound is heard, and then listens to the time-expanded sequence on a loop stored in the detector memory (Petersson detectors), either recording or discarding each sample in turn. Some detectors enable the observer to listen in heterodyne mode as well as record into either time expansion or frequency division.

The transect should be walked along a predefined route and at a steady speed. It can be useful to incorporate between 10 and 12 listening station stops interspersed along the chosen route. The length of each stop may be 2 to 5 minutes, depending on the overall length of the transect, and bearing in mind that it should take no more than 3 hours overall. Choosing easily identifiable, permanent features such as 'veteran oak tree', 'outside farm barn' as listening station stops ensures that the transect is repeatable by others, if necessary. Alternatively, the transect can be walked without stops, particularly when using broadband detectors.

Transects undertaken for conservation and monitoring are generally different to those undertaken for developments. A common objective in conservation and monitoring surveys, such as those undertaken for BCT's National Bat Monitoring Programme, is to observe changes over time (i.e. of population and species). To achieve this, the surveys are at randomly allocated sites, standardised and made repeatable. For monitoring purposes, it is important that each transect starts at a defined time, follows a set route and is undertaken at the same time of year.

When using transects to monitor bats for a development, the objective is to find not only an estimate of population and the species present but also all parts of the site which are used by bats. Therefore, it is an advantage to:

- undertake transects in reverse (or clockwise and anti-clockwise in the same survey);
- change the transect to look at a different area; and/or
- stop and start transects depending on the level or type of activity.

As for all activity surveys, walked transects are a combination of observation and listening. It is useful to develop a survey sheet (see Table 4.8 for an example) on which the species seen/heard, the time and the location (on or between listening station stops) can be noted. Alternatively, the comment facility available on some detectors can be used for this purpose.

Table 4.8 Example of walked transect survey sheet

Project name		Date		Sunset		Weather	
Surveyors			Bat passes heard				
Station number	Location	Time	Common pipistrelle	Soprano pipistrelle	*Myotis* or long-eared	Large bats (noctule, serotine, Leisler's)	Other
1							
2							
3							
4							
5							

4.6.2 Driven transects

Driven transects can cover much larger areas than walked ones. They can be one long transect or a series of short ones (where travel between the series records no data) and can also incorporate listening station stops. Driven transects may consist of more than one circuit of the route in the case of a small site to be surveyed, or partial transects undertaken over several nights, if the site is very large.

Preparation

Using aerial photographs, an Ordnance Survey map and a daytime visit, a route should be chosen that will incorporate habitat features with potential for use by foraging and commuting bats. It is recommended that an orange rotating beacon be mounted on the vehicle roof as a warning to other road users and that a sign be placed on the vehicle saying "SURVEYING". It is also recommended that the authorities (for example, police and local councils) be informed of the survey.

Method

The transect is driven along a predefined route and at a steady speed 15 mph (24 kph), continually recording bat sounds with a detector mounted out of the window or sunroof on the hedgerow side of the vehicle at a 45° angle. A time expansion or frequency division detector is normally used for the recording and this is subsequently analysed using sonogram analysis software. The use of time expansion bat detectors allows for the best chance of identifying bats to species level. The location of bat contacts can also be estimated from a simultaneous

GPS/Sat Nav record of the transect route. During the survey, the ambient air temperature, humidity, cloud cover, wind speed and direction are recorded. On a plan of the transect route, the locations of street lighting (including bulb type e.g. sodium or metal halide) are recorded. Lux levels may also be recorded. The transect should preferably be driven with dipped headlights.

A speed of 15 mph will still allow for a recording to be interpreted down to species level. At higher speeds (e.g. 30 mph), interpretation of the recording is difficult and unreliable; however, such surveys can be used to simply detect the presence of bats.

This transect methodology can cover a large area with relatively few surveyors and does not require highly trained individuals. Analysis of the recording requires considerable expertise and time needs to be allocated for this task.

Interpreting the results should take into account that there is a reduced chance of recording the same bat twice when compared to walked transects; however, it is still possible that a single bat can be recorded more than once during a transect survey.

Where it is safe to do so, this transect method can be carried out on a bicycle.

4.6.3 Boat transects

Boat transects can also cover larger areas than walked ones and may permit access to a waterway that does not otherwise have easy access. It is recommended that the relevant authorities (for example, police and British Waterways) and land or river owners be informed of the survey.

Preparation

It must be ensured that the craft used is suitable for the waterway to be surveyed, that there is sufficient depth of water, and also adequate headroom beneath bridges or overhanging vegetation.

Method

The transect is conducted along the waterway, close to the bank (distance will depend on the type of waterway). The same detectors can be used as in walked or driven transects. The detector should be held or fixed angled downwards at 45° to the water at the side of the boat (i.e. towards the bankside). Alternatively, switch between horizontal and downwards, in order to detect bats flying near to or within bankside vegetation as well as those flying close to the water's surface.

As for the driven transect, GPS can be used to record the location of bat sightings, and ambient temperature, humidity, cloud cover, wind speed and direction should be recorded. It is worth noting that close to the bank ambient light levels may be greatly influenced by light spill from buildings or from towpath lighting. Light levels can be recorded using a lux meter.

4.6.4 Backtracking to find roosts

Preparation

The approach to locating roosts using manual field surveys was first developed in The Netherlands and is commonly referred to as 'backtracking'. The technique is based on four principles:

1. The earlier the bat is seen at sunset or the later it is seen at sunrise, then the closer it is likely to be to its roost (the exact time depends on the species under study; see Table 4.1).
2. Bats fly away from their roost at sunset and surveyors should move towards flying bats to locate the roost.
3. At sunrise bats fly towards their roost and surveyors should move in the same direction as the bats to locate the roost.
4. At sunrise some bats species swarm at roost entrances for between about 10 and 90 minutes before entering.

Method

In the evening, surveyors should search for bats from 30 minutes before sunset, noting the time bats are encountered and the direction of flight: e.g. west commuting (arrows on a detailed plan with time and species are invaluable). This information is pooled from all the surveyors on to a map to identify potential commuting routes and possible roost sites.

Beginning 2½ to 2 hours before dawn surveyors should search again, this time for returning bats, starting with the potential flight routes identified the previous evening. Surveyors should be vigilant for a concentration of flight activity as bats return to their roosts.

Although this technique is biased towards early emerging species with loud echolocation calls and those which form large roosts, it is possible to locate roosts of any species using this method.

4.6.5 Dusk emergence and dawn re-entry surveys

Preparation

Emergence and re-entry surveys are the primary methods for locating roosts in trees, buildings or built structures, as bats are not always found by internal and external inspection surveys (e.g. if the bats roost in areas that cannot be searched and/or leave little or no visible trace).

An emergence survey can also give a reasonable estimate of the number of bats present. The tree, building or built structure should be inspected in daylight before the survey is undertaken, using binoculars where necessary, in order to assess the features, all potential exit locations and the number of surveyors required.

All the surveyors should be briefed of these findings and informed of the areas on the tree, building or built structure they are to watch. Sufficient surveyors should be used so that all aspects of the tree, building or built structure can be viewed at one time; for a simple tree or regular four-sided building two surveyors are adequate but it is better to have some surveillance overlap. Surveyors should count a set area to avoid double counting. In public places the surveyors should be able to see each other; additional surveyors may be required and fluorescent jackets are a useful aid to visibility.

Some bat species do not emerge until 1 hour or more after sunset and the use of night vision equipment and bat detectors should be considered to aid observation and possible identification of species.

Method

Surveyors are positioned so that all possible bat exits can be observed at one time and the line-of-sight should not exceed 50 m. The length of time to undertake emergence surveys is given in Table 4.5 and the period in the year during which they can be carried out is given in Table 3.3.

Observation of a tree/building for over two hours requires vigilance and concentration. The bats are more easily observed against a light background, such as the evening sky, and with a bat detector to assist in detecting bats.

Dawn surveys are particularly revealing as bats generally spend more time returning to their roost than emerging from it, giving the surveyor more time to see the bat and entrance location; the dawn light can also be more favourable. August is an especially good month to observe maternity roost re-entry, as young flyers can be inexperienced and are often highly visible when returning to the roost. It should be noted that, as with emergence times, different species also vary in the time that they return to the roost. For example, pipistrelles and noctules may return when light levels are relatively high compared to other species. Periods when female bats are lactating will also affect the pattern of emergence and returning at maternity colonies over the course of the night and morning.

To obtain a reasonable estimate of the number of bats present, all bat exits and re-entries are counted. There is usually more than one roost exit, so the number of bats present in a tree, building or built structure cannot be calculated until all the surveyors report their numbers at the end of the survey.

Chapter 5

Automated bat activity surveys

5.1 Introduction

Automated systems can be employed to achieve a greater level of survey intensity than with manual bat detector surveys, and for minimal extra survey effort. They can allow several sample points to be surveyed at the same time, providing more comparable results, or be used to provide a more flexible timetable for surveying. There are two types of systems that can be used to record bat activity remotely in the absence of a surveyor: remote bat detector recording systems and automated activity logging systems.

Remote bat detector recording systems consist of broadband detectors (frequency division or time expansion) attached to a recording device. Depending on the system used, this produces either a log of bat passes over time or records calls for later sound analysis, enabling an attempt to be made at species identification (see Table 4.2 for advantages and disadvantages of the different detector types).

Automated activity logging systems use a device triggered by sound (via a detector) or movement (crossing an infrared beam) and record these as events onto a data logger. Such systems have the potential to collect long runs of quantitative or semi-quantitative data on roost use, with minimal effort. The ideal automated activity logging system would count the number of bats going into and out of the site or derive an index of activity, identifying each bat to species and logging the time of each event. In reality, counting individual bats and identifying the species can be very challenging.

Night vision camcorders or cameras activated by movement can be used to record bat activity and possibly bat behaviour without the need for an observer to be present throughout the survey period. These systems are deployed in the field at specified sample points, for example, to observe a roost entrance.

Automated techniques of both bat detector recording and activity logging can be undertaken as an alternative to walkover surveys in some circumstances but do not provide observer notes, which are often essential to identify certain species. They are best considered as a cost-effective way of increasing sampling effort and ideally, both techniques would be used in a complementary manner.

For information on assessing the need for a bat survey and its planning and preparation refer to Chapters 2 and 3.

5.2 Equipment

Technology is changing rapidly and new systems are being developed that allow equipment to be left unattended in the field for long periods. Appendix 2 describes some systems that have been used regularly in the UK up to the time of publication but readers are advised to consult the relevant specialist suppliers for details of any new systems that are available or in development. Background information on the different types of systems available is also given in Glover and Altringham (2007).

Fairly simple systems can be put together with little technical knowledge, using commercially available components, but there can be advantages in commissioning specially designed systems. This is especially the case for large-scale surveys or where there are not enough observers available to provide the necessary cover within the given time or budget.

As they are left in the field unattended, automated systems have problems of security and power supply. They can often be costly to purchase and require a secure, vandal-proof site. It is not always possible to weatherproof the equipment completely nor to find a site that will remain dry. Equipment running on batteries may require regular battery changes or the use of a solar panel or other means of natural energy generation in order to charge the batteries.

A recent case, where a bat detector recording system placed on a bridge was the subject of a security alert and 'bomb' disposal, emphasises the need to advise the relevant authorities (e.g. police, Highways Authority) to prevent the destruction of expensive equipment and disruption to the public.

5.2.1 Bat detector recording systems

Complete systems that are designed to be left out in the field to detect and record bat calls are commercially available. It is also possible to build a system by combining standard bat detectors and recording devices that use tapes, mini-discs, memory cards or hard disk MP3 recorders. This option requires some electronics expertise but can provide a cheaper system or more units. This could be an advantage particularly for bat group or student projects, where funding is limited but skilled voluntary help may be available, or where there is a greater risk of the units being lost or damaged.

Commercial systems are more expensive to buy but there can be cost savings in the time taken to set up the system and to download and analyse the data. For example, one system can store many nights' recordings on one Compact Flash (CF) memory card and may include software that condenses the data, making it much quicker to scan for bat calls than using a separate recording device that contains bat calls interspersed with 'white noise'.

Some of the considerations that will influence any decision about the best system to use for a particular survey are:

- information required from the survey;
- number of sample points and therefore how many units are required;
- type of detectors to be used (see Table 4.2 for advantages and disadvantages of the different types);
- other methods available (e.g. to complement activity loggers);
- number and availability of surveyors;
- level of surveyor expertise (in the field, in electronic systems and in sound analysis); and
- security of site and protection from the elements.

Automated systems are relatively easy to set up and deploy in the field. Once the system is place, data can be downloaded and the battery changed by unskilled personnel. Expertise is required to decide how many units are required, where to locate them and for analysing the resulting data.

5.2.2 Automated activity logging systems

Automated activity logging systems use a device that is triggered by sound (via a detector) or movement (crossing an infrared beam) and record these as events onto a data logger or to a camera. Camcorders with a night vision function can also be used to record activity and, with the availability of built-in hard disk drive storage, increasingly have the capacity to record for long periods of time.

Automated activity logging systems differ from automated counting systems. Activity logging systems can be used to derive an index of activity at a location but are not accurate enough to count the number of bats. It is not possible to determine if the bats being detected or filmed are different individuals unless the equipment is being used at a roost entrance.

Systems that only log activity, but do not record bat calls, will need to be supplemented by further observations or recordings for identification purposes. However, even very simple recording systems can save time by determining which areas at a site merit more detailed survey work, or which time of the night or season should be the priority for more work.

The restrictions that apply to automated activity logging systems are the same as those encountered with the same sort of equipment in any other situation and many factors need to be considered when deciding which system to use. Some systems will be more appropriate for use with certain species or in particular locations, as some species are more easily detected than others and detectors vary in their sensitivity to different frequency ranges. These factors can be used to advantage in surveys concentrating on a particular species but are a restriction for surveys looking at the broad spectrum of bat activity.

5.3 Timing, survey effort and frequency

Activity loggers can be used at any time of year but low or negative results obtained when the weather is less favourable for bat activity need careful interpretation. During the summer, the system can be used to record bats as they commute, forage, or move to and from daytime and night-time roosts. During autumn, they can be used to provide a gauge of the level of swarming activity at a site (see Section 7.3.3). Bats arouse periodically from hibernation but as activity levels will be lower in winter, bats are most likely to be picked up close to a roost entrance.

Automated systems of both types can be used to supplement transect survey data by leaving the system running throughout the night when observers are carrying out other surveys nearby.

Using a system of remote detection and bat sound recording, the numbers of bat passes recorded can vary appreciably from night to night but the overall pattern of activity through the night, and the proportions of different species, are likely to be similar on successive nights. It is recommended that such systems be used in each location for three rain-free nights in succession, in order to give representative figures for that time of year. As a minimum, three separate sessions should be carried out at each location, spaced between May and September, in order to record seasonal variations in activity. See Table 4.7 for frequencies and timings for manual surveys - for sites surveyed both manually and remotely, an intermediate level of survey effort would be required.

Alternatively, programmable automated detectors can be left to record continuously from mid-May through to mid-September, switching on and off at pre-set times. The detector should activate 30 minutes before dusk and switch off at either midnight or dawn, depending on the survey design. The only constraints here are security, the weather and the need to change batteries or link the battery to a self-charging system such as a solar panel. For example, the equipment could be left to run indefinitely, with only infrequent visits to change batteries and download data, if the location being surveyed is secure and protected from rain, such as inside a barn or a grilled mine.

In order to correlate results to weather conditions, the use of temperature and humidity data loggers in conjunction with the automated systems is required.

5.4 Location of loggers and detectors

5.4.1. Automated detectors

Bat activity varies markedly with habitat, so there is a need to identify the different habitats and structures and to consider their use for roosting, foraging and as flight paths by different species. Before automated detectors are positioned, the purpose of the survey must be considered. A few examples of where to site the detector system in order to suit the survey purpose are given in Table 5.1.

The flight characteristics of the species under study and the location will need to be considered. A survey of high-flying bats in an open area would require different placement of detectors than a survey of activity of a bat species associated with cluttered habitats. The likelihood of detecting different groups of bats will need to be included in the interpretation of the results.

Automated systems placed in the open can supplement transect surveys and indicate whether the results are representative of activity at other parts of a large site.

Table 5.1 Examples of locations of bat detector systems according to survey purpose

Purpose of survey	Location of detector
To determine if a barn is used by bats.	On a suitable beam above head height in the barn.
To determine if an underground site is used by autumn swarming bats.	Inside the entrance of a grilled underground site.
To identify commuting activity along a hedge line to be severed by a road proposal.	In the hedge line at the position of severance.
To determine if a woodland ride is used for foraging.	On a tree facing into the ride.
To determine bat activity in the area of a proposed wind farm (see Box 5.1).	Detector microphone fixed to wind monitoring mast or balloon to simulate the height of the nacelle.

Box 5.1. Note on surveying for wind turbine developments

The significance of any effects of wind turbines on bats in the UK has not yet been determined. Research in the US and in other European countries indicates that wind turbines have a detrimental effect on some bat species, identified as tree-roosting bats, aerial-feeding bats and particularly on migratory bat species. Research is needed in the UK to determine which species are vulnerable, how significant the effect may be and how it may best be mitigated (for example, by changing the proposed location to avoid important bat areas). It should be remembered that development associated with wind farms, e.g. access roads, may have impacts on bats in addition to the turbines themselves.

The methodology for bat activity surveys at wind farm sites is still being developed. It is recommended that anyone undertaking or assessing surveys for proposed wind farm sites should consult the generic guidelines contained in Annex 1 to Resolution 5.6 of the Fifth Meeting of Parties to the EUROBATS Agreement **www.eurobats.org**

Resolution 4.7 of the Fourth Meeting of Parties to the EUROBATS Agreement (also available from **www.eurobats.org**) emphasises that, until research is completed, a precautionary principle should be adopted in relation to decisions on the development and siting of wind farms, especially along migration routes and in areas of particular value to bat populations.

5.4.2. Activity loggers

Automated activity logging is particularly effective where bats are constrained into a narrow flight path, such as under bridges or at roost entrances. If the location is secure, the equipment can be left in place over periods of days, weeks or even months. The use of such systems along flight paths or close to roost entrances can give an indication of the number of bats commuting to or from the roost, if the time of the observations is recorded. If several time-linked detectors are used, the direction of flight can also be deduced.

5.5 Sound analysis software

There are several different sound analysis software packages suitable for analysing bat recordings. Two packages specifically designed for the analysis of bat recordings are *BatSound* and *BatScan*; others are general-purpose sound analysis packages (e.g. *Spectrogram, Audition* [formerly *Cool Edit*] and *WaveSurfer*), which require more expertise if they are to be customised for bat recordings. The software *Analook* is designed specifically for use with the Anabat system. It does not analyse sound recordings directly but displays data from a sound wave that has already been processed in real-time by the hardware.

5.5.1 *Analook*

Software for the analysis of Anabat recordings is supplied with the system and kept updated by downloads from **www.hoarybat.com**. There is a full and documented DOS version of the software but most users now use *AnalookW*, a Windows-based version. *AnalookW* is still in development and is not yet documented but it is easy to learn and notes on its use are available.

The frequency-time displays of bat calls can be shown at a wide range of resolutions. The comprehensive software provides many analysis aids including the facility to label and annotate recordings. The results – species, time, location and other notes – can be extracted into text files for importing into spreadsheets in order to perform further analysis, chart plotting, etc.

5.5.2 *BatScan* and *BatSound*

BatScan and *BatSound* both accept sound recordings either played in via the computer sound card or loaded directly as .wav files. They allow the display of sonograms and power spectra of bat calls, which, together with the measurement of variables, such as pulse length and repetition rate, assist in accurately identifying bat species.

The software of both packages can be used with any frequency division or time expansion recordings, although the *BatScan* default settings have been optimised for use with the Batbox Duet bat detector. The choice of a software package will depend on cost, ease of use, additional analysis aids and the quality of the displays. Demonstration versions of the two packages are available but the opinions of existing users should also be sought.

Chapter 6

Surveying buildings and built structures

6.1 Introduction

All bat species resident in the UK have been recorded using buildings and built structures (e.g. bridges) at some time during the year. Many also use underground sites or trees for roosting; surveying these is dealt with in Chapters 7 and 8 respectively. Although this chapter deals mainly with surveys of buildings and built structures, much of the introduction is also of relevance to these later chapters because it describes roost characteristics and preferences of different species.

For the purposes of these guidelines the definitions of buildings and built structures outlined below have been used. These are broad definitions that should cover most circumstances encountered by bat surveyors.

Buildings

Buildings are also structures but they are characterised by having walls and a roof. Buildings include residential properties, flats, offices, warehouses, garden houses, follies, barns, stables, limekilns, towers, churches, former military pillboxes, schools, hospitals and village halls. Some areas of buildings, in particular cellars, have more features in common with underground sites.

Built structures

A built structure is something that has been man-made but is not a building or an underground site. The most commonly surveyed built structures for bats are bridges and walls but they can also be monuments, statues and chimneys. Within this category are also buildings that are in such a state of disrepair, having lost their roofs, that all that remains are the walls or part of the walls.

Some species, such as horseshoe bats, are found only in buildings or underground sites and not in trees. Others, such as the common pipistrelle and brown long-eared bat, are regularly found in a wide variety of built structures, buildings, underground sites and trees. Others, such as the barbastelle, rely heavily on trees and are only occasionally found in buildings but are regularly recorded in underground sites in winter.

Throughout the year, bats make use of differing environmental conditions in various types of roosts in order to meet the needs of their annual cycle (see Figure 3.2 for a diagram of this cycle). At its simplest, maternity roosts need to be warm, whereas winter hibernation sites need to be cool and humid. Bats occupy these two types of roost site for significant periods of time but in between these key periods of the year bats will make use of a wide range of other places for shelter and protection.

Bats are usually easier to locate where they occur in high numbers but will also be found individually or in small numbers in locations that cannot be observed directly or appear to be inaccessible to them. In these circumstances, surveying for bats can be very difficult.

As with most ecological survey work there is no substitute for practical experience. With greater experience comes an improved sense of likely roost sites within a building or built structure. However, experience can also lead to over-confidence, which can result in some potential roosting locations being dismissed before appropriate and thorough survey methods are applied.

This chapter aims to provide guidance on how to approach and undertake surveys of different types of buildings and built structures in order to determine the presence of bats. The guidance cannot cover all the possible circumstances that bat surveyors will encounter but is designed to be adapted to unique or unusual survey situations.

For information on assessing the need for a bat survey and its planning and preparation refer to Chapters 2 and 3.

6.1.1 Understanding roosts and making an assessment of the likelihood of bats being present

A basic understanding of the different types of roosts used by bats throughout the year is essential for bat surveyors, as it informs the initial assessment of the likelihood of bats being present and how a roost survey might be undertaken. General guidance on the preferences of different species for various types of roosts in the UK can be helpful and is provided in Table 6.1. This guidance is **not** definitive but it illustrates the wide range of places that bats can and do use.

Surveyors should also have a good knowledge of the basic biology of bats and the reasons for selection of

different roosts at various times of the year. Particularly important is roost use in relation to thermoregulation and energy conservation.

The use of a building or built structure by bats can sometimes be complex and the bat surveyor should always consider how the building or built structure could be used throughout the year. For example, consideration should be given to determining whether the building or built structure could be used for one or more of the following purposes:

- spring gathering roost that is used by breeding females before moving to a maternity roost;
- maternity or nursery roost where females give birth and raise their offspring;
- daytime summer roost used by males and/or non-breeding females;
- mating roost (spring or autumn) occupied by males seeking to attract females for breeding;
- night roosts used by bats for short periods between phases of foraging activity, but rarely or not used during the day;
- feeding roost or perch where bats temporarily hang up to devour an item of prey once it has been caught;
- transitional roost used for short periods in the spring and autumn;
- pre-emergence flight and foraging area either within the same building as the roost or in adjacent buildings. Such areas are used for a range of purposes including warming up, light testing before emerging, social interaction and cover from predation; and/or
- a hibernaculum.

For the purposes of this guidance, summer roost sites include those occupied from April to September although this also depends on the species, the geographical location and weather in any particular year. Some will be transitional, especially in the spring (March to May), allowing bats to become torpid during cold weather.

Transitional sites may also allow breeding bats to congregate before moving to the principal summer roost location. Other roosts in late summer and autumn include those that are formed by males seeking to attract females for mating.

Surveyors should also be aware that species vary in their use of summer roosts. For example, some species may use a main maternity roost continuously for several weeks, whilst others may use a number of sites in rotation, often returning to a site more than once during the breeding period.

Making an assessment of the likelihood of bats being present will require a level of expert judgement. For example, traditional farm buildings and many listed buildings, especially those close to good foraging habitat, are very likely to be used and will usually require a survey to determine the presence of bats. Other buildings or built structures, such as recently built modern industrial units or inner city structures, particularly if isolated from foraging habitats, are less likely to support bats.

Table 6.2 lists some of the features of buildings and built structures that surveyors should bear in mind when assessing the need for, or undertaking, a survey as they will influence the likelihood of bats, or particular species, being present. Further guidance on assessing the need for a survey is also provided in Section 2.3.

Table 6.1 Examples of places used by bats for shelter and protection

Species	Summer	Winter
Common and soprano pipistrelles	Often found in relatively modern houses (post-1940s), in confined spaces on external parts of buildings e.g. under lead flashing, in box eaves and cavity walls. Often found in buildings with flat roofs. Also found in older buildings and built structures such as walls and bridges.	Small crevices in buildings, trees, stone walls, bridges, barns and also in bat boxes. Often in fairly exposed locations to take advantage of warmer winter days for feeding. Rarely in caves and tunnels.
Nathusius' pipistrelle	Trees, buildings and bat boxes.	Trees, buildings and bat boxes.
Brown and grey long-eared bats	Older buildings with large uncluttered roof spaces. Roost along the ridge beam, in mortise joints, gable ends and around chimney breasts.	Buildings, caves, mines, tunnels and ice houses. Will roost in crevices as well as in the open.
Natterer's bat	Old, often stone buildings with large main roof beams, timber-framed buildings and trees. In buildings they like small crevices such as those between beams or in mortise joints.	Cool entrances of caves, mines buildings and other underground structures - even exposed rock face crevices. They will often jam themselves into crevices.
Whiskered/Brandt's bats	Found in a range of buildings, old and new, but do have a preference for older buildings with stone walls and slate roofs. Crevice dwellers found under ridge tiles, under slates, behind rafters, hanging tiles and roof boarding.	Caves and tunnels, but also some sites barely underground such as follies and limekilns. In caves they tend to occupy areas close to the entrance.
Daubenton's bat	Bridges, especially over water, tree cavities, mill-races, tunnels, mines and cellars. Occasionally in buildings, usually old stone ones. More frequently found within houses in Scotland than other parts of the UK.	Caves, mines, buildings and other underground sites. Often found in the warmer, more stable environments. They can be found closer to the entrance towards the end of winter.
Bechstein's bat	Woodland species. Roosts in trees and bat boxes.	Trees.
Noctule	Primarily a tree dweller. Very rarely in buildings and structures such as walls. Will use bat boxes.	Trees, rock fissures, hollows and bat boxes.
Leisler's bat	Tree holes, bat boxes and buildings, both old and new. Found around gable ends, under felt or ridge tiles, and under loft insulation. Highly mobile species and roosts can be occupied for only a few days.	Tree holes, crevices in buildings and occasionally in caves and other underground sites.
Serotine	Typically in 1930s buildings with high gable ends and cavity walls. Also occur in much older buildings and often in churches. Less frequent in modern buildings. Rarely obvious in the building. Found in cavity walls, crevices, around chimneys, under ridge tiles, between felt and roof tiles or boarding and roof tiles.	Thought that most hibernate in buildings.
Barbastelle	Trees and occasionally use of buildings such as timber-framed barns. Rarely found using buildings for breeding purposes.	Buildings, rarely in caves and tunnels.
Greater horseshoe bat	Buildings, particularly older ones with a large fly-in access point to an open roof. Sites include older manor houses, churches and barns.	Caves, disused mines and tunnels, some of which can be 50 km or more from the breeding roost.
Lesser horseshoe bat	Larger rural houses, barns and stable blocks offering a range of roof spaces and a nearby cellar, cave, tunnel or ice house where they can go torpid in poor weather. Prefer access with uninterrupted flight, but can use more inconspicuous gaps if necessary (e.g. under door gaps).	Caves, disused mines and tunnels, some of which can be 50 km or more from the breeding roost.

Table 6.2 Features of buildings and built structures that are considered to influence use by bats in summer[1] (adapted from Mitchell-Jones, 2004)

Likelihood of bats being present	Feature of the building or built structure and its location
Increased likelihood	Pre 20th century or early 20th century construction[2]. Agricultural buildings of traditional brick, stone or timber construction. Large and complicated roof void with unobstructed flying spaces. Large (>20 cm) roof timbers with mortise joints, cracks and holes. Entrances for bats to fly through. Poorly maintained fabric providing ready access points for bats into roofs, walls, bridges, but at the same time not being too draughty and cool. Roof warmed by the sun, in particular south facing roofs. Weatherboarding and/or hanging tiles with gaps. Undisturbed building roofs and structures. Bridge structures, follies, aqueducts and viaducts over water and/or wet ground. For rarer species, building or built structure is located in the core area of the distribution. Buildings and built structures in proximity to each other providing a variety of roosting opportunities throughout the year. Buildings or built structures close to good foraging habitat, in particular mature trees, parkland, woodland or wetland, especially in a rural setting (see Boxes 2.1 and 3.1).
Decreased likelihood	Modern, well maintained buildings[3] or built structures that provide few opportunities for access by bats. Small cluttered roof space. Buildings and built structures comprised primarily of prefabricated steel and sheet materials. Cool, shaded, light or draughty roof voids. Roof voids with a dense cover of cobwebs and no sections of clean ridge board. High level of regular disturbance. Highly urbanised location with few or no mature trees, parkland, woodland or wetland.

Notes

1 This table relates to assessing the potential use of buildings and built structures during the summer. It should be noted that these features may not necessarily be indicative of use by bats during winter or spring. This table should be read in conjunction with Box 2.1, which provides triggers for bat surveys.

2 Pre-1914 buildings may present the greatest likelihood of providing roost space for bats due to their design, materials used and ageing. Pre-1960 buildings, especially when close to good foraging habitat, and with favoured features such as cavity walls and soffits, can also have a high likelihood of providing roost sites for some bat species.

3 Post-1960 buildings are generally less likely than older buildings to house roosts; however, some modern designs can provide access to suitable roosting spaces for bats. Pipistrelle bats in particular will occupy modern buildings and built structures providing that there are suitable access gaps (> 8 mm) and that the building or built structure provides appropriate characteristics for roosting. Such situations should be carefully considered.

6.1.2 Aims of survey

A survey of a building or built structure should aim to determine:

- if bats are, or have been, present within the building or built structure and, if so, which species are present;
- the type of roost (e.g. maternity roost, day roost used by males or non-breeding females, feeding perch, night roost, mating roost, transitional roost, hibernaculum);
- how bats use the building or built structure (e.g. location of roosting bats, flight paths and flight behaviour, exit and entrance points to the roost); and
- the intensity of use (e.g. number of bats, sex of bats, time and duration of use).

6.2 Methods

The aims of the survey, the locations of the bats in the building or built structure and the time of year that the survey is being undertaken will all influence the method employed.

Survey methods appropriate for buildings and structures can broadly be categorised as:

- external and internal inspection survey;
- dusk emergence survey;
- dawn re-entry survey;
- netting and trapping survey; and
- automated activity survey.

Dusk emergence and dawn re-entry surveys are discussed in detail in Chapter 4 and automated surveys in Chapter 5. Methods of trapping bats are discussed in Chapter 9. Surveying for hibernacula at underground sites and autumn swarming surveys are considered in more detail in Chapter 7.

In this chapter, methods for undertaking inspection surveys of buildings and built structures are discussed in detail. The use of backtracking and emergence and dawn re-entry surveys, in terms of the effort that may be required and the approaches that may be employed, are also considered.

6.2.1 Approach to inspection surveys

Inspection surveys can be used to determine the actual or likely presence of bats and how they use a building or built structure. This includes the location of all known or likely roost sites and evidence of whether they are used by bats. For many built structures, such as bridges, walls or buildings where there are no roof voids, internal inspections are not practical and further survey work using different methods may be required. However, where possible, internal inspection of the building or built structure should be included.

When assessing a site's potential as a hibernaculum, surveyors should be aware that bats may hibernate in places that cannot be accessed through normal surveying methods and this may lower the confidence in a negative survey result.

To lawfully enter a known roost site, surveyors must be in possession of an appropriate SNCO licence or be accompanied by an appropriately licensed person. For details of surveyor licensing, see Section 3.4.

An inspection survey, whether internal or external, should be carried out as follows.

- Permissions should be obtained to enter the building and any land with a clear view of the building. Consideration should be given to any equipment (ladders, platforms etc.) required to allow the survey to be undertaken, especially if entering roof voids. Equipment that may be useful for inspection surveys is given in Box 6.1.

- An appropriate risk assessment should be completed following best practice such as that recommended by the Health and Safety Executive (see Section 3.5). Internal inspections of buildings in particular can expose the surveyor to a wide range of hazards, from falls to asbestos (see Section 3.5).

- Sufficient time should be allowed during daylight hours to enable a thorough survey to be completed of all external and internal parts of the building or built structure.

- If possible, information should be requested about sightings of bats from on-site personnel such as security guards, owners of the property or neighbours. However, it should be noted that the observer may not be objective or may lack experience, so anecdotal information should be treated with caution.

- The results of the survey should be recorded in a standard manner, for example, on a Bat Roost Report Form such as that given in Appendix 3 of the *Bat Workers' Manual* (Mitchell-Jones and McLeish, 2004) including, as a minimum, the information provided in Box 6.2. The location of evidence of bat usage or potential access or roost sites should be noted on technical elevation and base plan drawings if these are available; alternatively, drawings of the building or built structure should be made on which to record the evidence. Correct architectural terms should be used to describe the different parts of the building; a glossary of terms and diagrams showing the different structures of roof void is provided in Appendix 1 of the *Bat Workers' Manual* (Mitchell-Jones and McLeish, 2004).

- Photographs are often extremely helpful in showing the structure and condition of a building or for recording evidence of bat presence, and it is recommended that digital photographs are taken to support written records. Permission may be required from the building owner to take photographs and care should be taken to avoid causing disturbance to roosting bats. Where no additional disturbance is caused, photography without flash is permissible as an incidental part of licensed conservation or scientific work. Note that the use of flash photography in roosts or hibernacula requires separate licensing (for further details see Mitchell-Jones and McLeish, 2004).

- A search should be made for direct evidence of bat presence. A systematic search pattern should be developed in order to avoid missing parts of the building or built structure. During the survey, a search should be made for droppings, corpses, scratch marks, urine staining, grease marks and clean cobweb-free gaps around potential entrance points and crevice roost sites, including along the ridge beam. Listen for sounds of bats. Potential access points and roosting sites should be recorded even if there is no direct evidence of use by bats. It is recommended that samples of any droppings are collected for comparison with a reference collection or photographic reference material. The inspection should be thorough and a consistent search effort applied to all parts of the building or built structure. It should always be remembered that many buildings and built structures support more than one roost site and may be used by more than one species.

6.2.2 Equipment

A list of equipment for undertaking surveys in buildings or built structures is provided in Box 6.1. It is important that surveys are undertaken using the most appropriate equipment in order to provide the necessary health and safety protection and to minimise the risk of missing bat presence. A range of personal protective equipment (PPE) may be required; this will be determined by the risk assessment and working Method Statement prepared prior to the survey. See Section 3.5 for detailed information on risk assessments and health and safety.

A powerful torch is required to survey a dark roof void so that no evidence of bats is missed because of poor illumination. Similarly, a good pair of binoculars is required when inspecting buildings or built structures from the ground, in order to view potential features that may be used by bats. Without binoculars, there is an increased risk of failing to see small numbers of bat droppings stuck on walls or window glass on upper storeys.

Endoscopes and mirrors can be used to look behind features of a building and in mortise joints or cavity walls, which can provide greater certainty about the presence of bats. Endoscopes should be used with care - direct contact with bats must be avoided and the endoscope should be slowly and carefully manoeuvred. Furthermore, the close proximity of the light and heat from some endoscopes can cause significant disturbance to roosting bats so they should not be used for longer than is necessary.

Additional equipment will be needed for an emergence or re-entry survey; Box 6.1 gives details, and further advice on the use of bat detectors for activity surveys is provided in Chapter 4. Equipment for automated bat surveys is described in more detail in Chapter 5.

Box 6.1 Equipment for use during inspection surveys of buildings and built structures

Binoculars

Powerful torch to illuminate dark corners from the ground, preferably with a red filter to minimise disturbance to bats

Endoscope and mirrors for inspection behind boarding and in cavities

Small torches for close inspection of cavities and cracks

Spare batteries for torches

Ladders

Access platforms (e.g. scaffold)

Reporting form

Collection pots and labels for corpses and droppings

Elevation and baseline drawings of the building or structure

Camera to record evidence and potential roosting sites

Personal protective equipment (PPE)

Bat detector

Boat to inspect waterside structures such as retaining walls and bridges

Thermometer

Compass

Tape measure or equivalent equipment to measure or make an accurate estimate of the dimensions of the roof void

Clinometer to measure height of building or structure and roost emergence points

Two-way radio

Box 6.2 Standard information to be recorded in inspection surveys

Evidence of use by bats	Features of the building or built structure
Location and number of any live bats.	Type of building.
Location and number of any corpses or skeletons.	Age of building.
Location and number of droppings.	Aspect of building.
Notes on relative freshness, shape and size of droppings.	Wall construction, in particular the type of brick or stone used to build the wall and whether it has cavity or rubble-filled walls.
Location and quantity of feeding remains.	Form of the roof, in particular the presence of gable ends, hipped roofs, etc. and the nature and condition of the roof covering.
Location of clean, cobweb-free timbers, crevices and holes.	Presence of hanging tiles, weatherboarding or other forms of cladding.
Location of characteristic staining from urine and/or grease marks.	Nature of the eaves, in particular if they are sealed by a soffit or boxed eave and the tightness of the fit to the exterior walls.
Location of known and potential access points to the roost.	Presence and condition of lead flashing.
Location of the characteristic smell of bats if no other evidence is recorded	Gaps under eaves, around windows, under tiles, lead flashing, etc.
	Presence and type of roof lining.
	Presence of roof insulation.
	Presence of water tanks in loft (note if covered or uncovered).
	Structure of the roof including the truss type, age and nature of timber work.
	Information or evidence of work having been undertaken that could affect use of the structure by bats.

6.2.3 Searching buildings for bats

Species most often recorded from buildings include:

- common, soprano and Nathusius' pipistrelles;
- brown and grey long-eared bats;
- serotines;
- greater and lesser horseshoe bats;
- Natterer's bats;
- whiskered/Brandt's bats; and
- Daubenton's bats.

However, it should be noted that barbastelle, Leisler's, noctule and Bechstein's bats have also all been recorded in buildings and structures but these occurrences are rare compared to the species listed above.

For residential, commercial and stone-built agricultural buildings, the features that should be given particular attention during an external inspection survey include:

- holes in walls, pipes, gaps behind window frames, lintels and doorways;
- cracks and crevices in stonework and brickwork;
- gaps between ridge tiles and ridge and roof tiles, usually where the mortar has fallen out;
- gaps between lintels above doors and windows;
- broken or lifted roof tiles;
- lifted lead flashing around chimneys, dormer windows, roof valleys and ridges and hips or where lead flashing replaces tiles;
- gaps between the eaves, soffit board and outside walls;
- gaps behind weatherboarding, hanging tiles and fascia boarding;
- suitable entry and exit points around the eaves, soffits, fascia and barge boarding and under tiles;
- the presence of cavity walls and rubble-filled walls; and
- bat droppings on the ground, ledges, windows, sills or walls or urine on window sills.

A search should be made of the ground, especially below potential access points, windows sills, window panes, walls, hanging tiles, weatherboarding, lead flashing, eaves, behind peeling paintwork or surfacing materials and under tiles. For buildings constructed from stone rather than brick, particular attention should be paid to cracks and crevices that provide protection from the elements. Such features are known to be used by small numbers of bats throughout the summer period and occasionally maternity roosts have been recorded where access to rubble-filled walls is available.

Once the external inspection has been completed, an internal inspection should be undertaken. In the case of derelict or abandoned residential, institutional or office buildings, bats may be using rooms and other spaces within what would have been the living or working space of the building and each room should be surveyed for bat presence. Surveyors should work quietly and check the buildings in a systematic manner working upwards from the entrance (checking any cellar space last). On entering an individual level (or room) the places bats are most likely to be should be checked first; for example, if there are droppings under the ridge beam, the area above should be looked at and in open warehouses the darker areas should be surveyed first.

Within rooms in the buildings, searches should pay particular attention to:

- the floor and surfaces of furniture;
- behind pictures, posters, furniture, peeling paintwork, wallpaper, plaster and boarded-up windows;
- window shutters and curtains;
- wooden panelling;
- lintels above doors and windows; and
- clean swept floors.

Even where the building is still occupied, an internal inspection of the upper floors is necessary. Close inspection of window sills and window glass can provide additional information not clearly seen from an external survey from the ground.

Frequently-used roost locations within roofs include:

- top of gable end walls;
- top of ridge and hip beam and other roof beams;
- mortise joints;
- the junction of roof timbers especially where ridge and hip beams meet;
- top of chimney breasts;
- behind purlins; and
- between tiles and the roof lining.

A search of the roof void should be made with particular attention paid to:

- all beams for free-hanging bats;
- clean swept floors;
- droppings beneath the ridge and hip beams of the roof and junctions between the two;
- droppings, urine staining on and at the base of dividing walls, gable end walls and around chimney breasts;
- droppings, urine staining and corpses on, under or in materials or boxes stored in the roof;
- droppings beneath purlins;
- droppings and corpses beneath roof insulation;
- corpses at the base of walls and near wall plates at the base of rafters;
- corpses in uncovered water and header tanks or other containers in the roof;
- bat-fly (Nycteribiid) pupal cases[19];
- scratch marks and characteristic staining from fur oil on timber and walls;
- mortise joints and junctions between roof timbers and between timbers and walls;
- clean gaps and sections of ridge beam and other timber and walls within the roof;
- gaps between lintels above windows or doors;
- light gaps in the roof indicating access points to the outside;
- access to cavity or rubble-filled walls; and
- cool areas suitable for torpor or hibernation.

In addition to searching the roof and other parts of the building the surveyor should also listen for bats squeaking and chattering as this can often give away a roost location that is not visible.

6.2.4 Searching bridges for bats

Bridges require a particular range of knowledge and approach to surveying to properly assess the presence of bats. Bats regularly use a wide range of bridges, especially those that cross slow-flowing watercourses or that are close to good foraging habitat (Billington and Norman, 1997). Species most often recorded from bridges include:

- Daubenton's bats;
- Natterer's bats;
- pipistrelles;
- brown long-eared bats;
- whiskered/Brandt's bats; and
- occasionally lesser horseshoe bats that are known to roost in box girder bridges.

Bats roost in many different locations within old and new bridges. Features offering potential include holes, cracks and crevices leading to voids, particularly where there is clear access. Roosting locations in which bats have been recorded in bridges include:

- widening joints;
- expansion joints;
- gaps at the corner of buttresses;
- widening gaps (where the width of the bridge has been increased, forming a gap between the original and new structure);
- cracks/crevices (usually over 100 mm deep) between stonework and brickwork where mortar has fallen out (locations include the underside of the bridge span and spandrel, parapet and abutment walls);
- drainage pipes and ducts; and
- internal voids within box girder bridges.

Access is one of the most difficult issues to address when undertaking inspection surveys of bridges. There are significant hazards to the bat surveyor because bridges are usually associated with roads, watercourses or railway lines. Where the bridge is over water, a boat may be required in order to properly inspect the underside of the bridge. If the bridge is tall, lifting equipment or scaffolding may be required in order to undertake a detailed inspection of potential roost sites. If the bridge is over a working railway line, road or canal then consideration will need to be given to the specific requirements and associated risks of working in these environments and work will always need to be agreed with the appropriate operating authority.

19 A reference for the identification of bat-fly pupal cases is Hutson (1984).

Once safe and appropriate access has been secured, searches should be made of the ground (especially below potential access points) and an inspection made of key features and locations within the bridge structure that are likely to support bats.

As with external inspections of buildings, the structure should be surveyed using binoculars for the presence of bat droppings, clean gaps, urine staining, scratches, characteristic staining from fur oil, bat-fly (Nycteribiid) pupal cases and droppings. Note that bridges, like trees, can often conceal bats within features that are not visible from the ground.

Where access can be arranged, it is recommended that potential or likely roost sites should be inspected closely using an endoscope, torch or mirrors. Bridges and walls do not usually have accessible areas but where there are internal voids that can be accessed, they should also be surveyed, following the guidance provided for buildings.

Where there is high potential for bats to be present, but it has not been possible to confirm bat usage, dusk emergence and/or dawn re-entry surveys should be undertaken. It is also known that bridges are often used as night roosts and for mating and social behaviour. These potential uses should also be accommodated within the survey methodology.

It is important to remember that bats may use bridges at almost any time of the year and timing of any work must take into account bat activity at each site. It is not unusual for several bat species to utilise different parts of the same bridge.

6.2.5 Dusk emergence, dawn re-entry and automated surveys of buildings and built structures

Following external and/or internal inspection of a building or built structure, further survey work in the form of a dusk emergence, dawn re-entry or automated survey may be undertaken to provide additional information on an identified roost or to provide a reasonable level of confidence that bats are not present. All three survey approaches can be used to provide additional information during the summer period, but automated surveys in combination with inspection surveys are likely to be more effective in determining winter use. See Chapter 4 for information on dusk emergence and dawn re-entry surveys and Chapter 5 for automated surveys.

Dusk emergence, dawn re-entry or automated surveys may be required when:

- A roost has been identified and it is important to collect more information in order to make an assessment of the importance of the roost and the potential significance of any impacts on it. This may include the number of bats within the colony, determination of access points, species and flight paths to and from the roost.
- A comprehensive internal inspection survey is not possible because of restricted access but the building or built structure has features that have a reasonable likelihood of supporting bats.
- A comprehensive inspection survey is not possible because it has been undertaken at a sub-optimal time of year and there is the risk that evidence of bat usage may have been removed by weather, human activities or the presence of livestock.
- Inspection surveys have not ruled out the reasonable likelihood of a roost being present. This decision will be based on expert judgement but the guidance given in this document should assist with such decisions.

Night-time bat activity surveys, although not emergence surveys, can also be useful in identifying summer night-use within buildings.

Dawn re-entry surveys are useful in detecting summer roosts, particularly those used by small numbers of bats, which can be difficult to find especially if they are species that emerge late or are seldom detected with a bat detector. It may be easier to see bats re-entering rather than emerging, especially if they spend time 'swarming' around their roost entrance at dawn.

Where it proves difficult to identify a roost entrance for species such as brown long-eared bats which emerge late in dark conditions and which do not echolocate loudly, it may be more effective to locate a surveyor within the roof void of a building to watch where bats exit. However, this survey should be undertaken with extreme caution as the presence of a surveyor in the roof may affect the normal behaviour of the bats. Certainly, no such surveys of this type should be undertaken between mid-May and mid-July while the females are pregnant or have dependent young.

Automated surveys can be very useful in determining both summer and winter use of a building or built structure especially where other survey techniques are considered to be less effective. In this context, automated surveys have the advantage of being less intrusive than manual ones and can be left 'surveying' over a long period of time at little additional cost. This is particularly advantageous when determining winter use of a building, as bat activity will be irregular throughout the winter period.

Automated bat surveys have shown that buildings, especially agricultural buildings, can be used by bats in most winter months. For example, automated monitoring of bat activity within Paston Barn Special Area of Conservation (SAC) has recorded bat activity at regular intervals through the winter period (English Nature, 2003).

6.3 Timing

Where the evidence of bat usage is not exposed to the weather or human interference, surveys for summer roosts in buildings and built structures can be undertaken at any time, as the signs of roosting bats such as droppings, urine staining, bodies and bones should remain throughout the year. However, confirming the presence of bats in locations that are not physically accessible to the surveyor (e.g. cavity walls or beneath tiles or barge boarding) or where evidence is removed by rain and wind, is difficult if the bats are not active; in such cases, bats will not be regularly producing fresh evidence of their presence or being heard within the roost.

Even in situations where bats roost in the open, evidence may be removed by cleaners or attempts may be made to hide the presence of bats. Finding droppings in locations such as barns with earth floors and buildings used by livestock is difficult, as droppings may be difficult to differentiate from the substrate and/or may not accumulate. Therefore, if there is a reasonable likelihood of bats, an emergence and/or dawn survey is recommended in order to provide confidence in a negative survey result.

Emergence or dawn surveys should be undertaken during the period that bats are most active (usually April through to the end of September). The timing of the survey work should be determined by the surveyor to optimise the likelihood of recording bats.

It should be remembered that surveys might need to be undertaken in the spring or early summer and again in the autumn to investigate the use of buildings or built structures as spring transitional roosts or mating roosts. At these times of year, activity may be low due to cooler temperatures or inclement weather, such as high winds and strong rain. Survey days should be chosen carefully in order to maximise the likelihood of bat activity.

If it is considered that there is a possibility of a maternity roost being present within the building or built structure, the optimum time to undertake emergence or dawn re-entry surveys is judged to be from May to August inclusive.

Emergence surveys should begin at least 15 minutes before sunset and continue for 2 hours after sunset in order to take account of all species. For example, Natterer's, Daubenton's and brown long-eared bats often emerge well after dark. Some species, for example, serotines and pipistrelles, can emerge sooner so an earlier start time (30 minutes before sunset) may be required. See Table 4.5 for guidance on appropriate start times and survey durations.

Dawn re-entry surveys for maternity roosts are often most successful in August when young bats have only just begun to fly and their attempts to re-enter the roost are both obvious and of extended duration; but such surveys can be useful at any time during the summer. Dawn surveys should begin at least 90 minutes before sunrise, as *Myotis* species return to their roost sites much earlier than other species. See Table 4.5 for guidance on appropriate start times and survey durations.

If a maternity roost is known to be present in a building or built structure, access to the roost location during June and July should be avoided in order to prevent disturbance to lactating females. At this time an indication of the size of the roost and the species can be obtained using indirect survey methods such as counting and recording bats as they emerge from the roost.

Following an inspection survey, the surveyor should be able to assess if the built structure has potential to support bats outside the summer season. If the built structure has frost-free areas that can be surveyed either directly, or with endoscopes or mirrors, then winter or autumn surveys should also be undertaken between November and March with January and February being the optimum survey period (see Chapter 7). Consideration should be given to the potential level of disturbance that may be caused to hibernating bats by an endoscope and its use should be kept to a minimum.

Automated surveys in buildings or built structures during the winter period should be undertaken between November and February.

6.4 Survey effort and frequency

The level of effort and the frequency of survey required will vary depending on the type of building or built structure under consideration and the likelihood of bats being present. Features of buildings and built structures most likely to be used by bats are in Table 6.2, but it is not just these that should inform the decision to undertake a building or built structure survey. The context of a building or built structure, sometimes more then the physical form, will also influence the likelihood of bats being present and therefore the level of survey effort required. This is discussed further in Sections 2.3 and 3.7.

To illustrate how context can influence an assessment of the presence of bats, and thus the survey design, the following scenario is provided by way of example.

A small, modern timber and corrugated metal shed may not, on its own, appear to be a particularly likely place to find bats, especially if it supports few features that would increase the chances of bats being present. However, if such a building is located in a wooded valley with few other buildings nearby and is in the south-west of England, then its potential to be used by bats is likely to increase. This increased likelihood will require a greater survey effort than a similar building located in a less favourable place with little or no foraging and commuting habitat close by. In a poor context, such a building may only require a single survey visit but in a rich context, such as that described, three survey visits using more than one survey method may be required during the main period of activity for bats.

Guidance on the minimum survey effort and frequency recommended for buildings and built structures with differing potential to support bats is given below. For emergence surveys, further information is also provided in Chapter 4 and more specifically in Table 4.7.

6.4.1 General points on survey effort

Where surveys are required of large buildings or complexes of buildings, it is recommended that internal and external inspections and/or general activity surveys are completed in advance of emergence and dawn re-entry surveys to help focus survey effort at the most likely roost entrances. For large buildings it is recommended that dawn re-entry surveys should be undertaken as well as emergence surveys.

It should also be noted that the survey design may need to be significantly altered if it has not been possible to undertake an internal or adequate external inspection. For example, access into a roof void or building may be prevented by the presence of asbestos. Under these circumstances increased emergence and dawn re-entry survey effort should be made in terms of the number of survey visits and the number of surveyors employed on any one survey.

Automated surveys for winter activity within buildings or built structures with a reasonable to high likelihood of bats being present should be undertaken over a minimum of two weeks per survey. To provide confidence in a negative result, a minimum of two two-week survey sessions should be undertaken during the winter period, with one of these surveys between December and February.

6.4.2 Internal and external inspection surveys

The time required for internal and external inspection surveys will be determined by the number of surveyors and the complexity of the building or built structure being surveyed. Surveys of relatively simple buildings may be straightforward and quick to complete; however, it takes time to view and understand the roof structure of complex buildings or groups of buildings. Finding and accessing multiple entrance points adds time to the survey process. Using a greater number of surveyors should reduce the time required.

As a guide, typical lengths of time to carry out inspection surveys for three different scenarios using a single surveyor are given below:

- an internal inspection of the roof area of an unexceptional four bedroom domestic property could probably take 1 to 2 hours;
- an internal inspection survey of a traditional timber-framed farm building may take between 1/2 and 1 person day; and
- an internal inspection of a large complex building such as a former hospital or stately home with numerous roof voids and buildings may take as much as 3 to 5 person days to complete.

Inspection surveys for winter roosts can be more time consuming, as surveys using endoscopes and mirrors are often required in order to search for individual or small groups of bats hidden in crevices.

Time taken for daytime external inspection surveys will also vary according to the complexity of the building. Evidence may not remain after rain or wind, so weather and time of year will have a bearing on the level of confidence that an external inspection will give.

6.4.3 Dusk emergence, dawn re-entry and automated surveys

Determining an appropriate level of emergence and dawn re-entry survey effort in order to have confidence in a negative survey result is not always a straightforward decision. The effort required is a combination of the number of survey visits, the number of surveyors and automated methods employed, the time of season in which the survey is undertaken, and the weather at the time of survey. These factors may also be complicated by the variety of buildings and built structures involved. In all cases a level of expert judgement will be required, with such judgements becoming harder as the survey requirements and circumstances become more complicated. However, the guidance given below is considered to represent the minimum level of effort required in most circumstances. If, in the expert judgement of the surveyor, a lesser survey effort is considered appropriate, then a clear explanation should be presented with the findings of the survey.

If the evidence is clear enough, a single daytime survey visit may be adequate to determine if bats are, or have been, present. However, a negative survey result from a single daytime visit to a building with moderate or high potential for bats should be supplemented by additional dusk/dawn surveys. This is especially important if the daytime visit has been conducted outside the optimum period. A single daytime survey with a negative survey result is not considered adequate unless the building or built structure concerned provides no suitable roost sites (i.e. is of low potential, refer to Tables 6.1 and 6.2 for further information).

The likelihood of determining the presence of bats is increased if survey work for summer roosts is undertaken at the optimum time of year and a variety of survey methods are used.

If bat signs are not present and the building or built structure provides no suitable locations for roosting bats, then no further visits would normally be required. However, under these circumstances it is important that sufficient information is provided to give confidence in the assessment. It should be noted that a single summer visit will not reveal winter use of a building or built structure, although an assessment of the likely use in the winter may be possible.

If a building or built structure is considered to have a low likelihood of use by bats it is considered that one dusk emergence survey together with a dawn re-entry or automated survey is often sufficient to provide confidence in a negative survey result.

If a building or built structure is considered to have a moderate or high likelihood of use by bats, it is recommended that at least three emergence surveys and/or dawn re-entry or automated surveys per building or built structure should be undertaken during the summer period in order to give confidence in a negative survey result. At least one of these surveys should be a dawn re-entry survey or automated survey. At least two of the surveys should be completed between mid-May and August.

As well as ensuring that the survey includes a sufficient numbers of visits, the effort in terms of surveyors and/or automated survey methodologies should also be such that there is confidence in the negative survey result. This is perhaps one of the hardest elements of bat survey design on which to provide prescriptive guidance. In general, however, the number of surveyors or automated survey systems (see Chapter 5) used should be sufficient to cover all likely roost entrances throughout the survey time, in one survey visit or over a number of visits, in order to ensure that all likely roost entrances are observed at the frequency described above.

For example, for a single small simple building such as a bungalow it may be possible to undertake an adequate level of survey with one surveyor and one or more automated survey methods. However, the health and safety risks associated with lone working at night and the efficacy of an automated survey technique compared to a surveyor should be carefully assessed. Two surveyors would be preferable.

For larger buildings such as halls, manor houses and institutional complexes (schools/hospitals, etc.), a greater number of surveyors are likely to be required in order to provide an appropriate level of survey. Ensure that surveyors cover all aspects of a building or built structure. The number of surveyors and automated survey methodologies should be clearly described in survey reports illustrating the coverage achieved by the survey design.

For large buildings or complexes with a moderate to high likelihood of bats it may be necessary to increase the number of survey visits to more than three and/or increase the number of surveyors in order to provide sufficient survey coverage.

Chapter 7

Surveying underground sites

7.1 Introduction

Bats use underground sites for two main purposes, hibernation and swarming (see section 7.3). Some species also occasionally use underground sites for breeding.

For hibernation, bats require sites that are humid and cool, with stable temperatures and little disturbance. For more detailed information about roost choice, and the species that use underground sites, refer to Table 6.1. The duration of hibernation depends largely on prevailing weather conditions; bats that might otherwise hibernate above ground, in hollow trees or crevices in buildings, may enter underground sites during particularly cold spells. Similarly, bats move location within underground sites depending on their thermoregulatory needs, and may be visible or hidden from sight. For example, they may move from an exposed position near the entrance of a cave to a secluded location deep within a crevice, in response to a change in ambient temperature.

Some underground sites are used for swarming during the spring and, most importantly, the autumn. This behaviour, which is believed to have a social function, begins in early August, peaks in mid-August to mid-September and ends in October or November, depending on the species and geographic location. During this period, many bats may arrive at the site after dusk, stay for a few hours and then leave. Therefore, few bats may be present at the site during the day.

Daubenton's bats and greater and lesser horseshoe bats have been recorded breeding in underground sites in the UK. There is, though, the possibility of finding small numbers of other species using underground sites as day roosts during the summer; these are most likely to be males or non-breeding females. Underground sites may be used as night roosts at any time of year, especially if they are close to good foraging areas. However, the emphasis in this chapter is on surveying for hibernation and swarming sites, reflecting the particular importance of underground sites for these activities.

In these guidelines, the definition of underground sites encompasses man-made and natural structures that are entirely or partially underground. They include caves, souterrains, mines, ducts, tunnels, military bunkers, wells, and ice houses. Some underground structures beneath buildings, such as cellars, are more readily accessible to surveyors than caves and mines; they are often used as part of the building, for example, for storage, and may not require the same survey approach as for underground sites. Instead they may require an approach similar to that described in Chapter 6 for buildings and built structures.

For information on assessing the need for a bat survey and its planning and preparation, refer to Chapters 2 and 3.

7.2 Surveying underground sites for hibernacula

7.2.1 Understanding hibernacula and making an assessment of the likelihood of bats being present

It is essential that bat surveyors have a basic understanding of the different types of hibernacula and the reasons why they are used, as this informs an initial assessment of the likelihood of bats being present and how the survey should be undertaken. Surveyors should also understand the reasons for roost selection at different times of year, being able to relate this to biology, including thermoregulation and energy conservation.

In these guidelines, hibernacula include those occupied from October to March inclusive. The use of such sites varies depending on the species, sexual status, age, geographical location and weather. Some of these hibernacula are transitional, used when bats are moving between summer and winter sites. Transitional sites often provide opportunities for bats to come out of torpor and take advantage of warmer weather and available prey items.

The core period of occupancy for significant hibernacula is December to the end of February, when weather conditions are normally at their harshest. This period varies across the UK and may be longer as one moves northwards.

The most frequented hibernation sites provide high humidity and protection from freezing temperatures and wide temperature fluctuations but offer a variety of environmental conditions (such as those found in complex cave and mine systems) and are close to good foraging habitat. Conversely, bats are less likely to use sites, such as cellars or underground car parks in urban areas, that have low humidity, temperatures that fluctuate in line with external conditions and are not in close proximity to foraging areas.

The range of species and the number of bats present at a hibernaculum can vary. Some sites, such as cavity walls in buildings (see also Chapter 6), may be occupied by very small numbers of bats (fewer than five) and be completely inaccessible to surveyors. Other hibernacula, in particular underground sites, can support a variety of species in large numbers. Some species, such as horseshoe bats, often hang in open situations, whilst others will squeeze into cracks and fissures, or even beneath rubble or stones close to or on the ground. However, horseshoe bats may roost in cavities and other bats may roost in the open. Some species, in particular barbastelles and long-eared bats, can tolerate low roosting temperatures.

Assessing the potential of underground sites to support bats can be undertaken relatively easily where there is safe access. However, determining the potential bat usage of buildings or built structures by bats in the winter is much harder to achieve with a high degree of confidence (see Chapter 6).

Because of their roosting habits, assessing the importance of a site for horseshoe bats may be easier than for other species, where only a small percentage of the bats present at a site may actually be visible at the time of the survey. The results of winter surveys in underground sites where there are many opportunities for bats to conceal themselves should be interpreted with caution. For this reason, surveys during the autumn swarming season can be a useful tool to learn more about the variety and numbers of species that use a site, particularly those species that may easily be missed during winter hibernation counts (see Section 7.3).

7.2.2 Aims of survey

A winter roost survey of an underground site should aim to determine:

- if bats are, or have been, present within the underground site and, if so, what species;
- how bats use the underground site (e.g. favoured roosting locations);
- the intensity of use during winter (e.g. number of bats, time and duration of use); and
- the potential for use of the site by bats in other seasons (e.g. spring/autumn transitional roost, summer maternity roost or autumn swarming).

The presence of large collections of droppings may indicate a summer maternity roost but as a general rule winter surveys are not appropriate for determining use of a site during the summer and the site should also be surveyed in summer to determine if it is also used at this time.

7.2.3 Methods

Preparation

Surveyors entering known hibernacula must be in possession of an appropriate SNCO licence or be accompanied by an appropriately licensed person. For details of surveyor licensing see Section 3.4.

It is advisable to consult mining history organisations, the British Caving Association (BCA, **http://british-caving.org.uk/**) or local caving groups before undertaking visits to natural caves and abandoned mines. Such organisations frequently have important information about safety precautions on or in the site, its layout, history and details of any access agreements. For example, the BCA has a *Cave Conservation Code* (downloadable from their website). These groups may also be available to provide training or practical assistance for survey work, and may even have records of bat use at the site that could mean further disturbance could be avoided or reduced.

Each underground site has a particular set of hazards and appropriate training, including Confined Spaces training, may be required, along with specialist equipment (see Section 7.2.5). Entrances to underground sites may be above ground level, for example, in quarries or cliff faces, requiring surveyors to have climbing experience and specialist equipment. No underground site should be entered without conducting a thorough risk assessment, and ensuring that appropriate training has been undertaken and suitable equipment is available; see also Section 3.5.

Undertaking the survey

The site should be searched systematically from the entrance, with the locations of any bats seen marked on a map or plan of the site. A good torch is essential to light the way and to illuminate cracks, crevices and ceilings. Apart from horseshoe bats, which usually hang in open situations on walls and ceilings of hibernacula, many bat species are often under-recorded because of their crevice-dwelling habits.

Careful inspection for droppings or oil staining around cracks and crevices, including rock piles, may also yield evidence of use by bats. Activity loggers, which record bats' ultrasonic calls over periods of days or weeks, may also be used to provide a greater level of survey intensity with minimal additional effort (see Chapter 5).

The presence of any significant accumulations of droppings, *Nycteribiid* pupal cases[20], stained or marked areas should be recorded, as these may indicate the presence of large numbers of bats during the summer. Further visits during the summer may be required in such situations.

Handling hibernating bats has been shown to have a detrimental effect (Speakman *et al.*, 1991) and this must be avoided; for most purposes it is not necessary. Whilst it may not be possible to identify individual bats to species (particularly some of the *Myotis* bats), for many

20 A reference for the identification of bat-fly pupal cases is Hutson (1984).

surveys it is sufficient to be able to state that bats use the site and to identify them to genus.

Note that flash photography of hibernating bats should be avoided as the noise and light could disturb them; it would also require a licence.

7.2.4 Timing and frequency

The period during which bats will hibernate in any given winter is variable and will be influenced by a number of factors, such as the ambient temperature and humidity. Bats can use hibernacula anytime between October and March, depending on the weather (see also Table 3.3). The highest numbers of bats in underground hibernacula are usually found in January.

During the winter, bats move around to sites that present the optimum environmental conditions for their age, sex and body weight. Many species are only found in underground sites when the weather is particularly cold. Bats will periodically arouse to drink, as well as to feed if it is warm enough for insects to be active – for lesser horseshoe bats this has been shown to be at temperatures above or around 6 - 7° C. Arousal may also be triggered by disturbance because of increased levels of noise, light or heat – this may result from the presence of surveyors. The disturbance is not always obvious to the observer at the time, as bats do not necessarily arouse immediately. There is evidence that the longer the bats have been in a torpid state, the more sensitive they are to arousal stimuli (Thomas, 1995).

Because winter surveys may potentially disturb hibernating bats, visits should be limited to the minimum necessary to gain the required information. If it is necessary to assess the numbers of bats using a site, two visits are recommended, one in mid-January and one again in mid-February. BCT's National Bat Monitoring Programme hibernation surveys are carried out once in January and once in February. However, if there are very cold temperatures during March, a third visit at this time may also be considered, especially if this follows a warm winter when temperatures rarely fell below freezing. Visits should be carried out by a small number of people in order to keep disturbance to a minimum and to prevent increases in temperature.

For sites used by more than 10 bats, a single survey during cold weather in January or February has a high probability of detecting at least one bat and thereby confirming the presence of a hibernaculum. Absence is more difficult to demonstrate (see Section 7.2.1) and, in some cases, it may be prudent to assume that a suitable site in good habitat and close to other known roost sites will be used at some time by bats.

Timing surveys to coincide with a variety of winter weather conditions will provide greater confidence in the survey results. The probability of locating bats is influenced by the nature of the site as most species, except horseshoe bats, tend to conceal themselves in crevices if these are available.

7.2.5 Equipment

Safety equipment is essential when working in underground sites. Personal protective clothing should include gloves, warm clothing, protective headgear and footwear. A basic first aid kit should be carried and it is advisable to take food and drink when visiting large, complex or remote sites. Other specialist equipment, such as harnesses, ladders and/or ropes, radon monitoring badges, gas meters and breathing equipment, may be required before entering certain unfamiliar or more hazardous caves and mines, and training in the use of such equipment is vital. The BCA may be able to provide information about specific sites and on specialist equipment and procedures.

Equipment required for a winter hibernation survey includes the following:

- 'cool' torch (one producing as little heat as possible) for small sites;
- high power lamp for surveying large caverns;
- hard hat with head lamp;
- thermometer and humidity gauge to record ambient, entrance and internal temperatures and humidity;
- endoscope and/or mirrors for use in examining cracks and crevices;
- small pot to collect any droppings; and
- map or plan, notebook and pen/pencil to record findings.

7.3 Surveying underground sites for swarming activity

7.3.1 Understanding swarming sites

The species composition of swarming bats may be very different from that of hibernating bats found at the site, particularly in areas where most hibernating bats are horseshoes. *Myotis* species are most frequently recorded during swarming activity but may not have been recorded at the site during hibernation (Parsons *et al.*, 2003). Surveys for swarming bats can therefore reveal the importance of a site for a local or regional population that could be missed during hibernation surveys.

The importance of conserving a site that is used both for swarming and for hibernation cannot be over-emphasised. This is because swarming most likely serves a social function and is the mechanism whereby some bat species (in particular the *Myotis* species) locate mates from different colonies. This is thought to be important to improve the genetic mixing of the population and hence its reproductive health or 'fitness' (Rivers *et al.*, 2005). A single swarming site may serve a population from at least a 60 km radius around the site.

7.3.2 Aims of survey

A survey during the swarming season should aim to determine:

- if bats visit a site during the autumn (or spring) for swarming;
- the species or genus of bats present at the site (note it is not always necessary to identify bats to species level; it may be enough to know that large numbers of *Myotis* species are present); and
- the intensity of use of the site for swarming (e.g. an index of activity from automated detectors).

7.3.3 Methods

During swarming activity, bats enter and leave a site regularly during the night so the best place to look for swarming is at the site entrance. Surveyors entering a known bat roost (swarming sites are often also hibernation sites), or intending to catch bats, must be in possession of an appropriate SNCO licence.

Manual detectors, mist nets or harp traps can be used at the site entrance. Automated detector systems or activity loggers, which record or log bats' ultrasonic calls over periods of days or weeks, may be preferable because they provide a greater level of survey intensity and are less invasive than catching bats. However, these can usually only be used at secure sites, for example, where the equipment can be located behind a locked grille.

Identification of bats to species level is also constrained when using this method and a longer time will be spent on sound analysis (see Chapter 5).

Mist netting and harp trapping are highly invasive and labour intensive techniques and are more suitable for detailed scientific studies where identification of *Myotis* bats to species level is required.

7.3.4 Timing and frequency

Surveys for swarming bats are best carried out during August, September and October. Automated recording or logging systems can be left in place for the duration of the swarming season, or even longer, to provide a comparison of the level of activity during swarming with the rest of the year. Surveys for spring swarming can be carried out in March and April but this time of year is not preferable as activity is generally far lower than in the autumn.

Bat detector surveys and catching surveys must, of necessity, be less intensive than automated surveys, because of the number of people required and the potential for disturbance. One survey per month between August and October should be sufficient to identify whether a site is used by bats for swarming, provided relatively warm, calm and rain-free evenings are selected for surveying. Surveys should begin at 1 hour after sunset and continue for up to 4 hours.

7.3.5 Equipment

To undertake a survey of a swarming site, surveyors may require one or more of the following, depending on the protocol they determine to be most appropriate (in addition to the items given in Section 7.2.5):

- manual and/or automated bat detector systems (see Chapters 4 and 5); and/or
- mist net and/or harp traps and associated equipment (e.g. gloves and cloth bags) (see Chapter 9).

Chapter 8

Surveying trees

8.1 Introduction

All UK bat species rely on trees and woodlands to some extent; they provide valuable foraging areas, shelter from the weather and predators and represent key components of a connected landscape. However, perhaps the most significant use of trees by bats is as roosts and most species that occur in the UK use trees in this way. The dynamic ecology of trees and woodland provides a diversity of features that offer bats suitable conditions for roosting throughout the year. For more detailed information about roost choice, please refer to Section 6.1.

The likelihood of bats being present should be considered for all works, from individual tree pruning to large-scale forestry. Clearly, the survey for an individual tree is not the same as for a whole forest or woodland. However, there are common principles for identifying features of trees that are used by bats, including characteristics in the growth and decay of the tree and signs of activity left by bats. This chapter will focus on general principles to help guide arborists and woodland managers in good practice for their own specific situations, as well as to guide ecologists undertaking surveys or providing training and guidance for those undertaking works in trees and woodlands. It should provide arborists and foresters with the necessary guidance to: 1. undertake a preliminary assessment themselves, therefore minimising the risk of damage or disturbance; and 2. know when further assistance from an ecologist will be required.

Tree works fall into three broad categories, all of which have the potential to impact negatively on bats:

- ❍ Arboriculture ('tree surgery') – where the operation is concerned with a number of individual trees. The purpose of the work is to manage the trees for health, safety, and aesthetics.
- ❍ Forestry and woodland management – where felling and management works are concerned with predetermined areas as part of long-term management. This is usually for timber production although other purposes including heritage, amenity, game management, wildlife and recreation may also apply.
- ❍ Trees and woodlands as part of development projects – where the primary reason for the works is to facilitate development, for example, construction of new housing or roads. Works on trees in this context are driven by the needs of the development.

As each of these categories of works has distinct and differing impacts, it is appropriate to consider each the way in which the value of trees for bats should be assessed for each in turn.

Foresters and arborists undertaking their own preliminary bat assessments should have as a minimum basic awareness or training. Bat consultants/ecologists provide the most comprehensive bat surveys and will require specific experience of woodland or tree survey techniques. It is also important to have a basic understanding of tree biology and ecology when surveying trees because relatively minor growth characteristics can indicate the presence of decay cavities or splits that may not be immediately visible to the inexperienced observer.

Local bat groups who have practical experience and local bat knowledge may be willing to assist woodland managers with surveys. Grant aid may also be available to survey important bat populations or species for woodland management purposes.

Because bats are highly mobile, the impact of the management of trees and woodlands should be considered well in advance of planned works in order to enable a reasoned and thorough assessment of any impact to be made. Failure to do so could lead to significant delays and/or additional costs.

The long-term and planned nature of woodland management, which is often laid out in grant applications and/or design plans, makes it more realistic to manage bats at a landscape scale. This process is described in more detail in *Woodland Management for Bats* (Forestry Commission for England and Wales *et al.*, 2005), which is aimed at those involved in forestry and woodland management, and describes appropriate survey and management techniques.

The legal protection afforded to bats and their roosts is strict and good practice is aimed at minimising the risk of damage or destruction of tree roosts though it will never entirely eliminate it. If a tree roost is damaged, it is important that consultants, arborists and forest managers can demonstrate that good practice was followed. It must be remembered that roosts are protected even if the bats are not present at the time of any incident. Legal protection focuses on roost sites but good practice will also take account of foraging and commuting sites.

For further information on legislation refer to Chapter 1. For information on assessing the need for a bat survey and its planning and preparation refer to Chapters 2 and 3.

8.2. Methods

These guidelines are based on an approach that requires proportionately more survey effort as the value of the habitat for bats, and the impact of the loss of that habitat, increases. They are designed to ensure that woodland bat populations are sustainable, rather than to focus on the management of individual tree roosts. By using this approach, a workable assessment can be made of individual trees (those most likely to be tended to by an arborist), and a proportionate level of effort can be used to evaluate woodland and forest areas which are more likely to be managed by a forester, woodland manager or agent. Where removal of trees is planned due to development, it is recommended that each tree is assessed and recorded individually or in discrete groups so that survey and mitigation can be tailored to each feature of value. This is because the loss of a wooded area to development is, in the long-term, markedly different to felled areas within managed woodland.

The level of survey effort will be dictated by the amount of information required from the survey. The first stage is to establish the features of most value to bats. This includes individual trees with splits, holes or loose bark or groups of older trees in a woodland, where potential roost sites are more abundant. The value of a particular site for bats must be taken into account when the impact is considered; this is especially important for developments leading to significant tree loss.

Tree owners/managers should be reminded at an early stage that trees must not be removed prior to the granting of planning permission. It is important that bats are not harmed and/or flight lines lost, rendering existing roosts and planned mitigation features redundant.

8.2.1. Preliminary survey – all trees/woodlands

All trees near to work that is likely to have an impact on bats should have a preliminary visual inspection. Whilst this survey can be carried out at any time of year, winter surveys when there are no leaves on the trees will reveal more potential, while summer surveys will be more likely to reveal signs of activity. In some instances both types of survey may be required.

A preliminary survey should be undertaken as follows:

- Use close-focusing binoculars to inspect tree(s) from the ground to the canopy.
- Inspect all aspects of the tree(s).
- Look for features indicative of bat roosts (see Box 8.1). Use a high power torch, even in daylight, to inspect cavities and shaded areas of the branch structure. This is useful at any time but especially so on a dull day. On bright days, it may be necessary to time the surveys so as not to be dazzled by sunlight.
- For surveys undertaken in summer, listen for bats making audible social calls from roosts in trees - there is an example on the CD-ROM that accompanies *Woodland Management for Bats* (Forestry Commission for England and Wales *et al.*, 2005).
- Ask appropriate people (landowners, managers, wardens, workers or dog walkers) if there is a history of bats using the site; be aware this may not be accurate.
- Record findings on a map. It may also be useful to mark the tree with tape, paint or a tag.

Box 8.1 Common types of features used by bats for roosting and shelter and some field signs that may indicate use by bats (adapted from the *Arboriculture and Bats* training course ©BCT/Lantra Awards, 2005).

Features of trees used as bat roosts	Signs indicating possible use by bats
Natural holes	Tiny scratches around entry point.
Woodpecker holes	Staining around entry point.
Cracks/splits in major limbs	Bat droppings in/around/below entrance.
Loose bark	Audible squeaking at dusk or in warm weather.
Behind dense, thick-stemmed ivy	Flies around entry point.
Hollows/cavities	Distinctive smell of bats.
Within dense epicormic growth	Smoothing of surfaces around cavity.
Bird and bat boxes	

8.2.2 Assessing the value of trees affected by arboricultural works

Box 8.2 contains a suggested survey protocol for trees due to be affected by arboricultural works. This method of scoring a tree helps to relate the value of a feature to a recommended action. The specific criteria within the scoring system are less important than the outcome of each particular score; for example, High value = needs further inspection/survey while Low value = proceed with vigilance.

This approach should be considered by arborists as the basic standard for assessing trees prior to pruning or felling. For arboricultural operations, an inspection should be made at the earliest opportunity, for example, by an arborist when 'pricing-up' a job or undertaking a tree risk assessment. Failure to identify high potential features or signs of a bat roost could cause considerable delays, if these are detected later.

It should be emphasised that when signs of bat use are found, or where the potential for use remains high after closer inspection, then the assistance of an ecologist with appropriate experience should be sought.

Whilst the principle of this assessment method is transferable to all types of bat/tree evaluation, it is not considered appropriate for tree works on development sites (see instead Section 8.2.4), where more intensive survey work is likely to be required in order to assess the value of the trees along with other structures at the site, and to determine the impact on flight lines and from general disturbance over a longer period of time. Survey effort should be proportionate to the likely impact of any given operation.

Box 8.2 Bat survey protocol for trees due to be affected by arboricultural work
This would consist of a visual assessment of each tree for the likely presence of bats, according to the categories presented below. (Adapted from a protocol provided by SLR Consulting Limited)

Tree category and description	Stage 1 Survey requirements prior to determination	Stage 2 Further measures to inform mitigation	Stage 3 Likely mitigation
Category 1 Confirmed bat roost tree with field evidence of the presence of bats, e.g. droppings, scratch marks, grease marks or urine staining.	Tree identified on a map and on the ground. Further assessment to provide a best expert judgement on the likely use of the roost, numbers and species of bat, by analysis of droppings or other field evidence. **Ecologist involvement will be required.**	Avoid disturbance to trees where possible[1]. Further dusk and dawn surveys to establish more accurately the presence, species, numbers and type of roost present, and to inform the requirements for mitigation if felling is required.	Felled under Habitats Regulations licence[2] following the installation of equivalent habitats as a replacement. Felling would be undertaken taking reasonable avoidance measures[3] such as 'soft felling' to minimise the risk of harm to individual bats.
Category 2a Trees that have a high potential to support bat roosts	Tree identified on a map and on the ground. Further assessed to provide a best expert judgement on the potential use of suitable cavities, based on the habitat preferences of bats. **Ecologist involvement may be required.**	Avoid disturbance to trees where possible[1]. More detailed, off-the-ground visual assessment Further dusk and dawn surveys to establish the presence of bats and, if present, the species, numbers and type of roost to inform the requirements for mitigation if felling is required.	Trees with confirmed roosts following further survey would be upgraded to Category 1 and felled under licence as above. Trees with no confirmed roosts would be downgraded to Category 2b and felled taking reasonable avoidance measures[3].
Category 2b Trees with a moderate/low potential to support bat roosts	None. **Ecologist involvement is unlikely to be required.**	Avoid disturbance to trees where possible[1]. No further surveys.	Trees would be felled taking reasonable avoidance measures[3].
Category 3 Trees with negligible potential to support bat roosts	None. **Ecologist involvement will not be required unless new evidence is found.**	None.	No mitigation for bats required.

Notes

1 A general principle for those involved in advising on and undertaking tree works should be, wherever possible, to avoid disturbance and retain all features which offer some value to bats. For safety-related tree work, a balance should be sought between tree safety standards and the impact on wildlife.

2 When a Habitats Regulations licence to undertake work on a tree roost is required, the licence will need to demonstrate that alternative approaches have been previously considered to try to avoid works to the tree. These may be options such as diverting paths away from hazardous trees and removing unsafe limbs, instead of felling an entire tree.

3 Reasonable avoidance measures are considered to be good practice. 'Soft felling' is a generic term used to describe more cautious felling approaches, using lowering and cushioning techniques to reduce the impact of felling limbs which may still have bats within cavities. Where proportionate to the impact, best practice approaches to felling may include methods such as additional dusk emergence or dawn re-entry surveys immediately prior to felling (during the active bat season) or the use of non-return valves to ensure that bats can leave but not return to a roost cavity before works begin.

8.2.3 Assessing the value of trees affected by woodland management or forestry works

Woodland Management for Bats (Forestry Commission for England and Wales *et al.*, 2005) uses an approach that guides the surveyor towards locating areas of potential for bats, to signal if further survey is required, to locate actual roosts and to identify areas that should be left standing (sometimes known as 'Natural Reserves' or 'Minimum Intervention Areas'). Surveys should be phased, starting with areas due to be felled, rather than attempting to survey all forest blocks within a short timeframe, which would be very labour and time intensive.

The survey effort will be determined by the size, characteristics and age of the tree or woodland, its context in the landscape and the scale of the work proposed in relation to these features. For example, a brief walkover survey may be all that is required for the clearance of a young stand of plantation trees that is a small proportion of a large forest that has areas of old growth woodland. On the other hand, a stand of ancient or veteran trees should have a detailed inspection before any management works are carried out. An initial assessment of the area should be carried out and then survey effort concentrated in areas where a high likelihood of the presence of bats is indicated, preferably surveying twice, once in winter when deciduous trees are without leaves, and again in summer when recent signs or sounds of activity may be heard.

It must be emphasised that an experienced ecologist should be consulted if high potential trees, or actual signs of bat use, are found in areas scheduled for felling or thinning operations.

It is essential that records are made of all surveys to illustrate how the findings have been used to influence management. In addition, this helps to demonstrate that steps have been taken to avoid reckless disturbance or damage.

8.2.4 Assessing the value of trees on development sites

Tree works on development sites can have permanent or long-term impacts on the character of an area and may limit the opportunities for bats. Examples include loss of roosts (including maternity roosts, transitional roosts, mating roosts and hibernacula), loss of feeding habitat, impact on commuting routes and lack of shelter from weather and light. Where this type of work could impact on bats, it is essential that trees on and around the development site are assessed comprehensively for all features of value. This information will be required to inform the overall mitigation strategy.

Other bat survey work on the site may inevitably have some overlap with the survey effort required specifically for the trees. Consequently, suggestions for gauging the amount of survey effort required are provided in Section 8.4. Where larger areas of woodland may be affected by a development, but not removed altogether, it is recommended that appropriate advice from *Woodland Management for Bats* (Forestry Commission for England and Wales *et al.*, 2005) is used in conjunction with other mitigation measures, for example, identifying rich habitat areas and maintaining minimum intervention areas to ensure long-term habitat availability.

8.2.5 Preliminary survey area coverage

All parts of the woodland, or all of the trees in and around a development or management area, should be inspected. This may require prior arrangement with landowners, managers or on-site security personnel. A plan of the woodland or site should be used to guide the survey and record the findings. The area to be covered will be informed by the initial scoping exercise, and will be influenced by factors such as the species present in the locality, the connectivity of a site and the presence of known roosts. See Sections 3.6 to 3.8 for more information on desk studies, walkover surveys and scoping exercises.

A particular challenge when surveying trees is to keep the amount of survey effort proportionate to the likely impact of the works being carried out. It is essential that arborists, foresters and other land managers should be able to undertake their own initial assessment with limited knowledge and training, calling in specialists only where the evidence or likelihood of bat presence justifies it.

The inspection of trees, looking for features that could be used by bats, is similar to surveys undertaken for arboricultural reasons. For instance, a tree risk assessment will identify decay cavities and branch splits, so existing assessment procedures may be adapted in order to consider the features in relation to their value as bat roosts. Note, however, that these two functions may conflict, as bats often use the features that make a tree hazardous! This can be addressed by training of the personnel involved.

8.2.6 Detailed survey

Where the value of an individual tree or woodland area is found to be high during a preliminary survey, it is necessary to undertake a more comprehensive survey of the bats present in order to assess the impact of any proposed management. Should the scope of a potential impact change, for example, an increase in the number of trees to be felled, then further assessment and survey, proportionate to the level of impact, may be necessary.

The scope of additional work will vary considerably, depending on factors such as the habitat type, geographic location and altitude.

More advanced survey techniques are likely to include:

- climb-and-inspect assessment of selected trees to check for cavities and inspection with an endoscope or mirrors if appropriate;

- dusk and dawn surveys to observe bats leaving and returning to roosts (see Chapter 4);
- back-tracking surveys to follow bats back to roosts at dawn (see Chapter 4); and/or
- automated bat detector surveys to identify commuting routes and the frequency of bat passes at certain times (see Chapter 5).

A health and safety risk assessment should always be carried out and safe working methods should be adopted. Climb-and-inspect surveys should be undertaken only by suitably trained personnel; the use of ropes in trees, and even using ladders, carries specific risks (see Section 3.5).

Endoscopes can cause disturbance or even harm to bats and there are also limitations of endoscope length. See Section 6.2.2 for further discussion on the use of endoscopes in bat survey work.

All signs of use or potential use by bats should be recorded in a standard manner on a site plan or map. If bat droppings are found within or around a tree hole, it is recommended that a sample is collected for comparison with a reference collection.

Pre- and post-parturition (before and after birth) emergence counts may help to confirm whether breeding colonies are present. Exceptionally, capture may be necessary to confirm species, sex and breeding status or to fit radio-transmitters for further survey work (see Chapters 9 and 10), but these methods must be fully justified.

Examples of where advanced survey techniques such as mist netting, harp trapping or radio-tracking may be required include:

- a proposal with potentially major impacts on a bat population, for example, the complete loss of mature semi-natural woodland;
- a proposal with potential impact on important areas of habitat likely to support woodland specialist bats such as barbastelles or Bechstein's bats; or
- a proposal with potential impact on a designated area (SSSI, SAC, etc.), where bats form part of the special interest.

All groups of trees or woodlands exhibiting features of high value to bats should be subject to more detailed surveys involving night time and roost locating work. This becomes particularly important where planned felling will affect older (over 80 years) trees. Where woodlands meet the habitat requirements and distribution criteria for barbastelles and Bechstein's bats, specialist advice should be sought, and liaison with the appropriate SNCO is recommended (see Box 8.3).

In some circumstances, where other survey methods are considered insufficient or inappropriate, it may be appropriate to undertake precautionary exclusion using one-way flaps over holes and splits. This helps to ensure that no bats remain hidden within the tree when it is felled. However, the legal and licensing implications of this should be carefully considered to ensure that the law is not broken.

Box 8.3 Barbastelles and Bechstein's bats

Barbastelles may be found throughout most of lowland England and most parts of Wales. They roost primarily in trees all year round, although occasionally use old buildings such as barns. They normally select ancient or old growth deciduous woodlands with substantial understorey situated on the headwaters of small undeveloped catchment areas. They move roost site frequently and, in consequence, a large number of damaged and dead trees are normally present in favoured woodlands. Roost sites are chiefly in splits or behind loose bark.

Bechstein's bat nursery colonies can be found in central southern England and the southern Welsh borders. Nursery colonies are located in old growth or ancient deciduous woodlands of about 50 ha, although smaller blocks of suitable woodlands may be used, if they are well connected. There appears to be a strong tendency to select oak and ash woodlands. Small streams are normally present in these woodlands. Old woodpecker nest holes are the most frequently selected roost sites.

Should the presence of either of these species be suspected, it is advised that help is sought from ecologists experienced in these highly specialised species, and advice sought from the relevant SNCO.

8.3 Timing

Preliminary visual inspection surveys are best conducted during winter when trees are leafless and features are more visible but they can be conducted at other times of year if necessary.

The principles of timing for emergence and activity surveys are common to all survey scenarios. For details refer to Chapter 4. It should be borne in mind that, although bats in woodlands often emerge early, there are some species that frequently emerge later in the evening. Consequently, night vision equipment will be required to achieve accurate counts and to identify emergence points.

Although bats do fly occasionally in winter, the chance of an emergence survey taking place on a night when bats are active is small. Therefore a climb-and-inspect survey may give a better indication of the presence or likely absence of bats, if the cavity can be examined to its full extent.

Dawn surveys are useful for detecting roosts in trees, particularly when used by small numbers of bats. These roosts can be difficult to find, especially if they are used by species that emerge late or echolocate quietly and are seldom heard on a bat detector. It is usually easier to see where bats enter at dawn, rather than where they emerge from at dusk. Many bat species spend time at dawn swarming around the roost entrance before entering, so the entrance points can be easier to locate. For details of dawn surveys refer to Chapter 4.

For details on the timing of catching and radio-tracking surveys in woodland, refer to Chapters 9 and 10 respectively.

8.4 Frequency

In a woodland context, it is not possible to state the number of surveys that should be undertaken for a given site. The survey must ensure full coverage of all areas identified as having potential roosting sites and should take place in suitable weather conditions. If the development has a defined footprint, it is probably useful to compare the opportunities offered by the woodland to those offered by the buildings, and target survey effort accordingly. For example, a veteran oak tree could be compared to a small barn with thick walls and multiple cracks within timber mortise joints; or a group of 50 mature deciduous trees could be compared to a large Victorian school building with a complex roof structure. This is clearly a subjective judgement so it is essential that the amount of survey effort, and the frequency with which it is repeated, is justified in relation to known local bat species and bat flight activity around affected areas.

As tree-roosting bats can move roosts frequently, repeating the survey will improve the chance of finding roosts. However, even very intensive surveys will only ever find a small proportion of the roosts present in woodland, so the emphasis should be on identifying and managing high value areas. Survey results should be taken as a guide, together with other indicators such as trees with the potential to develop into future roost sites, to gain an understanding of how bats use the habitat.

It is best practice to assume that high potential trees are used by bats at some point during the year or will be in the future. Mitigation measures can then be provided accordingly, even if bat use cannot be confirmed and/or the operation licensed.

8.5 Equipment

The following equipment is required for a daytime assessment of trees:

- binoculars - to see into the crown;
- bright lamp - which helps to reduce shadow and can reveal true cavities;
- endoscope - may be useful, depending on the type of survey;
- notebook, site map or plan - to record findings; and
- access and inspection equipment (ladders or climbing equipment) - needed for climb-and-inspect surveys.

Torches and bat detectors are required for nocturnal surveys both at dusk and dawn. Night vision equipment (video camera or scope) is essential for effective emergence surveys of tree roosts; see Chapter 4 for details. For details of the equipment required for catching or radio-tracking bats, see Chapters 9 and 10.

Chapter 9

Catching surveys

9.1 Introduction

Mist netting and the use of harp traps to catch bats are well-established research methods and, as with radio-tracking, they can provide a large amount of detailed information in a relatively short space of time. However, because these are invasive methods, it should be considered carefully if catching is warranted and the protocol designed to minimise disturbance. The survey should catch the minimum number of bats necessary, and use the least invasive method appropriate, for the aims to be achieved.

Where building or land use developments are concerned, the need for catching bats should be based on evidence from comprehensive bat detector surveys for bat activity. The evidence should demonstrate that the species and numbers of bats and the predicted impacts from the development warrant this intrusive method of surveying to allow better impact assessment and mitigation design. It would be wise to discuss survey requirements with SNCOs before undertaking such work.

Bats may occasionally need to be caught for the purpose of:

- species identification (where it cannot confidently be done by a less invasive method - e.g. species that are difficult to survey using bat detectors, such as brown long-eared, grey long-eared and Bechstein's bats);
- sex determination (capture of females may indicate possible presence of a maternity roost);
- ascertaining breeding status; or
- attaching a radio transmitter (if radio-tracking is a necessary further step in survey, see Chapter 10).

Catching surveys are recommended if it is likely that proposed work will have a significant impact on bat populations. For example, if the pre-survey assessment or preliminary survey identifies potential for Bechstein's bats, then a catching survey could be undertaken in order to confirm the species' presence. It is considered likely that catching surveys would be required for development projects affecting woodland with high potential for bats, or for works affecting SACs designated for their bat interest.

A specific project licence is normally required for the capture of bats, and a sufficient level of expertise in capture, handling and identification techniques is required. Methods of capture must be specified on a licence application, including the justification for using an ultrasonic lure, if used, as this may cause additional stress to the bats. Catching surveys should be undertaken by specialist surveyors with prior demonstrable experience of the methods to be used (see the *Bat Workers' Manual* [Mitchell-Jones and McLeish, 2004]).

For information on assessing the need for a bat survey and its planning and preparation (including surveyor licensing) refer to Chapters 2 and 3.

9.2 Methods

Bats may be caught in hand nets or cone traps when emerging from roosts with small entrances, for example, building or tree roosts. For some species, such as pipistrelles and serotines, a hand net may be placed directly over the entrance. Some species, such as brown long-eared bats, can be difficult to catch from outside the roost and these may need to be caught by allowing them to drop into the net from their hanging place within the roost. This method has also been successfully used with greater horseshoes, Natterer's and Daubenton's bats but it must be carefully considered whether the greater level of disturbance inherent with entering the roost is warranted. Harp traps can be used at roosts when bats are exiting from multiple points, e.g. pipistrelles emerging from beneath hanging tiles or weather boarding.

When free flying, bats may be caught with mist nets or harp traps but success will depend on the level of bat activity and the careful siting of the nets or traps. Bat detectors should be used to identify commuting routes. Mist nets are then placed across the routes at dusk in areas where surrounding and overhanging vegetation funnels them towards the nets. This approach has worked well for some species, such as barbastelles, greater and lesser horseshoe bats.

Both mist nets and harp traps can be used with an ultrasonic lure that is designed to attract bats and increase the likelihood and rate of their capture. This is a relatively new technique which should be used with caution by persons with appropriate training and only where its use can be justified.

Mist nets and harp traps may also be sited at large roost entrances, for example, the opening of an underground site, to survey for swarming bats (see Section 7.3). An ultrasonic lure would not be necessary in this situation as bats are naturally funnelled toward the net or trap by the site entrance, and its use may potentially deter bats from entering the site. Mist netting is not advisable for use outside other sorts of roosts as, if many bats are caught in a short period, extraction from the net may take a long time and this may result in injury or death. In such locations, harp trapping is preferable.

9.3 Timing

The timing of surveys that incorporate the capture of bats will depend on the type of roost or habitat being studied. When catching at roosts, the survey should be timed to avoid catching heavily pregnant females, those that have recently given birth and are lactating, or those that may be carrying dependent young. In addition, avoid catching newly flying young.

In the field, mist nets or harp traps can be used at any time of night but will be most successful in the first few hours after dusk on calm, warm nights during the core activity period (see also Table 3.3). As at roosts, it is best practice not to catch in foraging habitats or on commuting routes when females may be heavily pregnant, newly lactating or carrying dependent young. It is difficult to give an exact window during which capture should be avoided, as the time at which bats give birth will vary according to the species, latitude and prevailing weather (births will be earlier when there has been an early spring). As a loose guide, the period from the end of May to mid-July is best avoided, unless there are exceptional circumstances where the survey specifically requires capture during these times.

9.4 Survey effort and frequency

To identify a species at a roost, capture of one bat on one occasion should be sufficient. This may also be the case when trying to identify flying bats. However, it may be necessary to capture more than one bat, for example, to determine sex ratios at a roost or where a minimum number is required for radio-tracking. It should be noted that the ratio of male to female bats at roosts can change throughout the year and that this has been particularly recorded at underground sites.

The minimum number of visits necessary is situation-dependent and also depends on the objectives of the survey. A clear picture of the information required from the survey should determine the effort required, and when sufficient information has been gathered, work should cease.

As an example, on a commuting route already identified as being used by bats through bat detector surveys, three nights of catching should be sufficient to catch bats for radio-tracking in order to locate a roost. If the objective is to identify species and/or sex, then one night in good conditions (not windy or raining) in each active bat month (May-October inclusive) should suffice, especially if several traps are used per km^2.

Capture probability can be enhanced by careful selection of survey sites - see the *Bat Workers' Manual* (Mitchell-Jones and McLeish, 2004) for further details. Visit frequency should be reduced where an ultrasonic lure is used, as this increases the capture probability but also the potential disturbance. For some species, the use of a lure is more effective in late summer/autumn than at other times of the year.

9.5 Equipment

For more detailed information on methods and equipment required for catching bats, refer to the *Bat Workers' Manual* (Mitchell-Jones and McLeish, 2004) and Kunz (1988). Information contained in the EUROBATS Resolution 4.6 (as amended by Resolution 5.5)[21] may also be of use. In brief, the equipment that may be required to catch bats includes the following:

- Hand nets – static butterfly hand nets are used to capture bats at the roost entrance or inside the roost.
- Cone traps – a net or trap adapted for use at the roost entrance to prevent bats from climbing out. The net may be left in place for a length of time to capture a number of bats.
- Mist nets – similar to nets used to catch birds, placed so as to intercept bats in flight. Can be used as bats fly away from the roost, on flight lines within foraging areas or on commuting routes.
- Harp traps – the costliest of the catching methods but less disturbing to bats than mist nets because the captured bats do not become tangled and stressed while being removed. Skill is required to set the tension of the harp trap strings correctly.
- Ultrasonic lures, e.g. Sussex Autobat[22], have proven effective at attracting some bats to mist nets and harp traps and hence increasing the capture rate. This method is most appropriate for survey in woodland where bats of several species cannot confidently be separated from one another by bat detector and where capture probability of those species that seldom use woodland rides is otherwise poor. The use of lures may also reduce by-catch of other species when a species-specific call is used. It should not be used at or near (within 50 m) roosts as it can induce high stress levels in some bats and could cause a roost to be abandoned.

21 EUROBATS Agreement Resolution 4.6 *Guidelines for the Issue of Permits for the Capture and Study of Captured Wild Bats* is contained within the written record of the Fourth Meeting of the Parties to the Agreement. Resolution 4.6 was subsequently amended by Resolution 5.5 at the Fifth Meeting of the Parties to the Agreement. Both meeting records are available online from **www.eurobats.org**. The UK is a Party to the Agreement.

22 The Sussex Autobat was developed by David Hill and Frank Greenaway. For further details see Hill and Greenaway, 2005.

Chapter 10

Radio-tracking surveys

10.1 Introduction

Radio-tracking is a powerful survey method for locating roosts, particularly of tree-dwelling bats that are otherwise difficult to find. It is also useful for determining the types of foraging areas and commuting routes used by bats from a particular roost or if the bats have alternative roosts nearby. Bats can be caught at, or close to, the roost in feeding or commuting areas, fitted with miniature radio-transmitters, and then tracked as they move to, and between, foraging areas or other roosts.

Radio-tracking is an invasive technique that should be used in situations where detailed knowledge is required about a population of bats. Such a technique is unlikely to be necessary for the majority of development cases but may be required for a landscape scale development where Habitats Directive Annex II species or other uncommon bat species are present, or when work may affect a SAC or SSSI designated for its bat interest. This level of survey will not normally be required for general woodland management unless it will have a high impact and high significance, for example, if Annex II species could be present. The need for such surveys should be discussed with the planning authority and SNCO before commissioning any work.

It should be remembered that the individual bats that are radio-tracked comprise only a sample of the whole population and, although detailed information is gained about their actual habitat use, they give only an indication of the type of habitat that the population as a whole uses. The remainder of the population may use other features that the radio-tracked bats did not use. This is especially the case for foraging areas, as individual bats usually have discrete foraging areas. Bats that are not radio-tracked may use different commuting routes to reach different foraging areas from those in the radio-tracked sample and the surveyor and client should be mindful of this in interpreting the survey results.

The objectives of a radio-tracking survey are likely to include:

- determination of detailed species composition and abundance;
- understanding of social and/or community structure;
- delineation of home ranges and the ecological function of the landscape for the species present;
- locating multiple woodland roosts;
- delineation of long distance flightlines; or
- recording the breeding success of subdivided colonies.

A survey should provide detailed information on roost, foraging habitat and commuting route use within the landscape. It may be less useful in terms of estimating how bats may use a specific site intended for development, particularly if the bats largely use areas outside the site. The aim should be to place the site in the context of the overall landscape used by, and the features important to, the bat population.

The objectives of the survey and the justification for the proposed level of radio-tracking should be clearly stated at the outset and subject to peer review if necessary.

A licence is required for the capture and fitting of bats with radio-transmitters. This type of survey should be undertaken by specialist surveyors with demonstrable radio-tracking experience. A sufficient level of expertise in capture, handling and radio-transmitter attachment techniques is required.

For information on assessing the need for a bat survey and its planning and preparation (including surveyor licensing) refer to Chapters 2 and 3.

10.2 Methods

For specific information on capture techniques, and the methods and best practice in attaching radio-transmitters to bats, refer to the *Bat Workers' Manual* (Mitchell-Jones and McLeish, 2004) and Kunz (1988). Information contained in EUROBATS Resolution 4.6 (as amended by Resolution 5.5)[23] may also be of use.

When bats have been radio-tagged, surveyors should aim to locate accurately all roosts of the tagged animals. These guidelines do not attempt to give details of the methodology for locating radio-tagged bats in the field. Reference should be made to other sources (e.g. Kenward, 2000) and knowledge is best gained through field experience. Once roosts are located by finding a radio-tagged bat, population sizes and breeding success may be estimated by emergence counts from the roosts (see Chapter 4).

23 EUROBATS Agreement Resolution 4.6 *Guidelines for the Issue of Permits for the Capture and Study of Captured Wild Bats* is contained within the written record of the Fourth Meeting of the Parties to the Agreement. Resolution 4.6 was subsequently amended by Resolution 5.5 at the Fifth Meeting of the Parties to the Agreement. Both meeting records are available online from **www.eurobats.org**. The UK is a Party to the Agreement.

Surveyors should also aim to identify key foraging areas and commuting routes. This can be achieved by the close approach method, where a single bat is followed continuously, or the triangulation method, where two receivers are used to 'fix' the bat's position. The latter method is less accurate but allows more bats to be tracked simultaneously. The level of resources required to collect these data can be substantial and, if the objective is to collect foraging and home range data, this needs full consideration in order to justify the tagging of multiple bats. The locations of roosts, foraging areas and commuting routes can be combined to represent the home range of a colony.

Careful consideration should be given to the number of bats used for radio-tracking and the benefits this brings in terms of data compared to the welfare considerations for the bats. In general, more effective results will be achieved with a higher sample size but it is recommended that catching of bats to enable radio-tracking should be kept to the minimum necessary to achieve the objectives of the study. The sample size should be sufficient to gain an overview of the whole colony without having to radio-track every bat. See Section 10.3 for further information on sample size.

Note that bats can also be marked with chemi-luminescent tags in order to observe hunting behaviour. The tags are short-lived, remaining visible for only a matter of hours. For further information, refer to the *Bat Workers' Manual* (Mitchell-Jones and McLeish, 2004).

Note also that ringing of bats is not recommended for survey. If several bouts of radio-tracking are anticipated, it may be appropriate to place rings on the bats that are tagged to ensure the same bats are not subsequently tagged. As individual bats suffer differently from the stress of being radio-tracked, tagging a bat for a second time risks inducing unacceptable stress and should be avoided. However, unless there is a specific monitoring objective or a long-term programme of research, ringing is not recommended or covered further by these guidelines.

10.3 Survey effort, timing and frequency

Radio-tracking of one bat of the most appropriate reproductive status and at the most appropriate time of year should be used to locate a colony's roost(s), bearing in mind that many species change roost every few nights.

Radio-tracking of a sample of bats should be used to determine foraging areas, commuting routes and home range of a colony (see Box 10.1 for examples). Depending on species, colony size and the intended use of the information, 5-30% of the colony (exceptionally 50%) would need to be tracked to have a working knowledge of the colony territory; 10-15% of the colony may be sufficient in a colony of 100 or less individuals and 5-10% in a colony numbering more than 100. It should not be necessary to track more than a maximum of 10-15 bats for any one colony.

In the event that the colony size is not known, for example, where the roost has to be located first, this judgement may only be made once an emergence count has been conducted to assess the colony size.

In addition, the number of bats to be tracked may be informed during the survey as the individuals' home ranges are delineated. When additional data do not significantly increase the colony home range (for example, when expressed as a Minimum Convex Polygon), then it is likely that enough bats have been tracked in that season. Work on lesser horseshoe colonies has found that approximately 10 bats were sufficient. Further disturbance by catching and radio-tracking additional bats can be avoided.

Box 10.1 Examples of radio-tracking survey protocols

Example 1. Intended clearfell of a woodland block within the territory of a known colony of Bechstein's bats. Objective is to understand how important that woodland block is to the colony and assess the impacts of the intended clearfell. Track 30% of the colony at low intensity.

Example 2. New road with street lighting cutting across a landscape with probable bat flight lines. All female bats suspected of being from different colonies trapped on the road line are tagged. These animals are followed to locate the roost(s). Then initially 5% of the colony is trapped and tagged from the roost. Further tagging may be necessary if it appears that a major problem exists, with up to 30% or more from rare species and important colonies. An alternative would be to obtain a broader picture of use by targeting bats using different flight paths or night roosts.

Example 3. Locate an alternative roost when a roost building is to be demolished. One bat is tagged for the life of the tag and located each day. When the tag fails it is replaced on a new animal until the colony moves and the new roost is located.

Radio-tracking twice in the year, once in May and again from late July and the end of August, will give a representation of the behaviour of a colony in different seasons. The core breeding period of early June to late July, when bats are heavily pregnant or have recently given birth and are lactating, should always be avoided unless absolutely necessary. In September, bats often range further than in summer and use more roosts, including swarming and hibernation sites. If underground sites occur within 10 km of a survey site, tracking may be required in this month as well.

The maximum survey effort is the lifetime of the radio-transmitter. Given the cost of the radio-transmitter and the invasive nature of radio-tracking, it is important to gain as much information as possible while the transmitter is active, and to make this available in the form of research reports or scientific papers in order to further understanding of the species. Ideally, radio-tracking should continue until no 'new' behaviour is observed. This may be 3, 4, 7 or more nights but extra nights may be necessary, for example, if bad weather causes a bat to remain torpid for one or more nights. The information gained should inform the duration for which tracking continues.

Minimum survey effort should be three full nights of tracking for each bat. This should be sufficient to meet the aims of roost location where bats use the same roost repeatedly, or to gain information solely along a specific flight path. This does not include the first night, as the bat may behave atypically immediately after tagging and release. In fact, the data gathered on the first night after tagging may not be representative and may be discarded at the surveyor's discretion. If the aims have not been met after three nights, radio-tracking should continue, especially when tracking tree-dwelling bats that may change roosts regularly.

To identify foraging areas, commuting routes and home ranges, a minimum of five full nights for each bat is preferable. It should become clear during the course of the radio-tracking survey whether representative data have been collected. For example, if after several nights no additional foraging areas or commuting routes are located and the home range increases no further.

It is possible to radio-track multiple bats in the same vicinity at the same time, especially if aided by a scanning receiver that enables ease of switching between radio frequencies. Alternatively, several radio-tracking teams with separate equipment may be required. This may be beneficial where a colony of tree-dwelling bats may be spread among several tree roosts.

10.4 Equipment

Specialist radio-tracking equipment is required. In brief, the pieces of equipment required for radio-tracking (in addition to those needed to capture the bats – see Chapter 9) are as follows:

- radio-transmitters and their means of attachment (surgical adhesive[24]);
- hand-held radio-receivers and antennas (those mounted on the roof of a vehicle may also be useful);
- vehicular transport; and
- compass, maps and GPS.

It may also be necessary to source computer programmes for the analysis of radio-tracking data, for example, to enable overlay on habitat maps and calculation of home ranges.

Kenward (2000) and Kunz (1988) are useful sources of further information on equipment, methods and analysis of data collected through radio-tracking.

24 Note that Skin-Bond® surgical cement, recommended for use in the *Bat Workers' Manual* (Mitchell-Jones and McLeish, 2004) has changed its formulation and is no longer suitable for use. An alternative, recommended by Holohil Systems Ltd. (see **www.holohil.com/bd2att.htm**), is Torbot Bonding Cement.

References

References marked with an asterisk (*) are not cited in the text but are recommended as useful background reading.

*Altringham, J.D. (1996) *Bats: Biology and Behaviour.* Oxford University Press, Oxford.

Altringham, J. D. (2003) *British Bats*. The New Naturalist Library, Volume 93. Harper Collins, London.

Barataud, M. (1996) *The Inaudible World* (2 CD set with booklet). Edition Sitelle.

Billington, G. E. and Norman, G. M. (1997) *The Conservation of Bats in Bridges Project - A Report on the Survey and Conservation of Bat Roosts in Cumbria.* COBIB/English Nature.

Boye, P. and Dietz, M. (2005) *Research Report No 661: Development of Good Practice Guidelines for Woodland Management for Bats.* English Nature, Peterborough.

British Standards Institution (2006) *PAS 2010 Planning to Halt the Loss of Biodiversity: Biodiversity Conservation Standards for Planning in the UK - Code of Practice.* BSI, London.

*Corbet, G.B. and Harris, S. (1991) *Handbook of British Mammals.* Blackwell Science, Oxford.

English Nature (2003) *Paston Great Barn Management Plan April 2003-March 2008.* English Nature Norfolk Team.

European Commission (2001) *Assessment of Plans and Projects Significantly Affecting Natura 2000 Sites - Methodological Guidance on the Provisions of Article 6 (3) and (4) of the Habitats Directive.* EC, Luxembourg.

Forestry Commission for England and Wales, Bat Conservation Trust, Countryside Council for Wales and English Nature (2005) *Woodland Management for Bats.* Forestry Commission, Wetherby.

Glover, A and Altringham, J. (2007) *A Review of Automated Bat Counting Systems.* Countryside Council for Wales Contract Science Report. CCW, Bangor.

Hill, D.A. and Greenaway, F. (2005) Effectiveness of an Acoustic Lure for Surveying Bats in British Woodlands. *Mammal Review* **35**: 116-122.

*Hill, J.E. and Smith, J.M. (1984) *Bats: A Natural History.* British Museum (Natural History), London.

Hutson, A.M. (1984) *Keds, Flat-flies and Bat-flies. Diptera, Hippoboscidae and Nycteribiidae.* Handbooks for the Identification of British Insects, 10 (7). Royal Entomological Society, London.

Hutterer, R., Ivanova, T., Meyer-Cords, C. and Rodrigues, L. (2005) *Bat Migrations in Europe. A Review of Banding Data and Literature.* Naturschutz und BiologischeViefalt 28. Federal Agency for Nature Conservation, Bonn.

Institute of Ecology and Environmental Management (2006) *Guidelines for Ecological Impact Assessment in the United Kingdom.* IEEM, Winchester.

Institute of Environmental Assessment (1995) *Guidelines for Baseline Ecological Assessment.* IEA, UK.

Joint Nature Conservation Committee (1989) *Guidelines for Selection of Biological SSSIs.* JNCC, Peterborough.

Joint Nature Conservation Committee (2004) *Handbook for Phase 1 Habitat Survey - A Technique for Environmental Audit.* JNCC, Peterborough.

Joint Nature Conservation Committee (2006) *National Vegetation Classification: Users' Handbook.* JNCC, Peterborough.

Jones K.E. and Walsh, A. (2001) *A Guide to British bats.* Field Studies Council / Mammal Society.

Kenward, R. E. (2000). *A Manual for Wildlife Radio Tagging (2nd Edn).* Academic Press. London.

Kunz, T.H. (Ed.) (1988) *Ecological and Behavioural Methods for the Study of Bats.* Smithsonian Institution Press, Washington, DC.

*Kunz, T.H. and Fenton, M.B. (Eds) (2003) *Bat Ecology.* University of Chicago Press, London.

*Kunz, T.H. and Racey, P.A. (Eds) (1998) *Bat Biology and Conservation.* Smithsonian Institution Press, Washington DC and London.

Limpens, H.G.J.A. (2005) *Vleermuizen en Planologie.* Zoogdiervereniging VZZ/Eco Consult and Project Management. The Netherlands.

Mitchell-Jones, A.J. (2004) *Bat Mitigation Guidelines.* English Nature, Peterborough.

Mitchell-Jones, A.J. and McLeish, A.P. (2004) *Bat Workers' Manual.* 3rd Edn. Joint Nature Conservation Committee, Peterborough.

*Mitchell-Jones, A.J., Amori, G., Bogdanowicz, W., Kryštufek, B., Reijnders, P.J.H., Spitzenberger, F., Stubbe, M., Thissen, J.B.M., Vohralík, V. and Ziman, J. (1999) *The Atlas of European Mammals.* T and A D Poyser, London.

*Neuweiler, G. (2000) *The Biology of Bats.* Oxford University Press, Oxford.

*Nowak, R.M. (1994) *Walker's Bats of the World.* John Hopkins University Press, Baltimore and London.

Parsons, K.N., Jones, G., Davidson-Watts, I. and Greenaway, F. (2003) Swarming of bats at underground sites in Britain - implications for conservation. *Biological Conservation* **111**: 63-70.

*Racey, P.A. and Swift, S.M. (Eds) (1995) *Ecology, Evolution and Behaviour of Bats.* Clarendon Press, Oxford.

*Ransome, R. (1990) *The Natural History of Hibernating Bats.* Christopher Helm, Kent.

Richardson, P. (2000) *Distribution Atlas of Bats in Britain and Ireland.* Bat Conservation Trust, London.

*Richardson, P. (2000) *Bats.* Whittet Books, London.

*Richardson, P. (2002) *Bats.* Natural History Museum Life Series. Natural History Museum, London.

Rivers, N.M., Butler R. K. and Altringham, J. D. (2005) Genetic population structure of Natterer's bats explained by mating at swarming sites and philopatry. *Molecular Ecology* **14**: 4299-4312.

Russ, J. (1999) *The Bats of Britain and Ireland. Echolocation Calls, Sound Analysis and Species Identification.* Alana Books, Bishop's Castle.

Russo, D. and Jones, G. (2002) Identification of twenty-two bat species (Mammalia:Chiroptera) from Italy by analysis of time-expanded recordings of echolocation calls. *Journal of Zoology, London* **258**: 91-103.

*Schober, W and Grimmberger, E. (1997) *The Bats of Europe and North America. Knowing Them, Identifying Them, Protecting Them.* TFH Publications.

Schofield, H. W. and Mitchell-Jones A. J. (2003) *The Bats of Britain and Ireland.* Vincent Wildlife Trust, London.

Speakman, J.R., Webb, P.I. and Racey, P.A (1991) Effects of disturbance on the energy expenditure of hibernating bats. *Journal of Applied Ecology* **28**: 1087 – 1104.

*Swift, S.M. (1998) *Long-Eared Bats.* T and AD Poyser, London.

Thomas, D.W. (1995) Hibernating bats are sensitive to nontactile human disturbance. *Journal of Mammalogy* **76**(3): 940 -946.

Tupinier, Y. (1997) *European Bats: Their World of Sound.* Edition Sitelle.

Appendix 1

Identification of bats by echolocation and flight

This Appendix contains information to supplement that given in Chapter 4. It outlines distinguishing features of echolocation calls and flight patterns to assist in identifying British bats, along with a brief methodology for sonogram analysis and references on bat identification.

Echolocation calls and flight patterns

The horseshoe bats are perhaps the easiest to identify to species level, having calls that are both different from vesper bats (all other UK bats are of the family Vespertilionidae hence 'vesper'; greater and lesser horseshoe bats belong to the family Rhinolophidae), and different from each other. Their calls are very directional so, despite being easy to identify, they can be easily missed. This is particularly the case with lesser horseshoe bats. Greater horseshoe bats prefer rough pasture and woodland edges, and often feed near wet habitats. They fly slowly with a fluttering flight and short glides, often flying low, even picking up prey from the ground, or ambushing prey from a perch. Lesser horseshoes hunt over similar ground and are also found in open and closed woodland, with shrub and canopy layers. Lesser horseshoes fly fairly fast, with fast wing movements. This bat species also flies close to the ground; rarely above 5 m. Lesser horseshoes often use a night perch to eat their prey.

Noctules and Leisler's bats can be distinguished by their call in an open environment and noctules can be heard over 200 m away. However, both adapt their call to the degree of clutter around them and in a cluttered environment are more or less indistinguishable. It helps to see the bat in flight and sound analysis can help distinguish between the species if the bats are flying in the open. Both noctules and Leisler's bats emerge early, and can be seen flying high, chasing insects, often at the same time as swifts. Noctules tend to fly higher and make fast turns and steeper dives than Leisler's bats. These species are not associated with linear commuting routes as are many other bats. Instead, they circle over woodland, woodland edge and pasture or above water. They can also be found foraging around white lights. The noctule is a larger bat but both have narrow, tapered wings which almost touch beneath the body as they fly. Serotines are almost as large as noctules, but they fly more slowly in large loops at about 6-10 m above the ground with occasional dives. Serotines prefer to hunt over old pasture, parkland, hedgerows and woodland edges, but can be seen in gardens and around lights. Serotines have broad wings when compared to noctules or Leisler's bats and this can make them look bigger, especially when seen from below. Serotines and Leisler's bats are often very difficult to distinguish from each other based on echolocation calls alone, particularly when they are in a cluttered habitat. Noctules and serotines also show overlap in call parameters in cluttered habitats. Even with sound analysis these species can be difficult to separate especially without additional information on habitat type and behaviour.

Although it is relatively easy to recognise *Myotis* bats from other species, the five members of this genus currently found in the UK (Daubenton's, Natterer's, whiskered, Brandt's and Bechstein's bats [there is only a single record of greater mouse-eared bat]) are the hardest to separate from each other in the field. They all use similar echolocation calls and the more additional field information gathered the better. *Myotis* bats, like all other bats, vary their calls according to the habitat in which they are flying, and call parameters overlap most when the bat is flying in a cluttered habitat, such as woodland. Sound analysis can assist in the identification of these species; however, it is not sufficiently robust in differentiating whiskered, Brandt's and Bechstein's bats solely from a sonogram. In acoustically cluttered conditions especially, call parameter overlap is often too high to attain the degree of certainty required. Moreover, sound analysis on its own is probably not sufficiently robust to differentiate any *Myotis* bats when they are outside their typical environment; for example, Daubenton's bats in woodland and Natterer's bats over a meadow.

Daubenton's bats are a medium-sized species, taking insects from close to the water surface or directly from the water itself, using their large feet as gaffs or the tail membrane as a scoop. They have a steady flight, at about 25 kph, often within a few centimetres of the water surface. They can also be found in woodland and along hedgerows and are more problematic to identify both by echolocation call and flight behaviour when foraging amongst vegetation.

In contrast, Natterer's bats fly higher over water, about 1 m over the surface. They will use smaller bodies of water, with thickly vegetated edges, and hunt over less calm water than that preferred by Daubenton's bats. Natterer's bats have a tighter 'turning circle' and change direction frequently. They do not tend to fly parallel to the water surface as do Daubenton's bats. Natterer's bats also have more abrupt changes in the call rhythm and their calls are quieter. They also fly in woodland, around tree canopies and other confined spaces, again making tight turns. Natterer's bats emerge late, often flying close to the roost entrance without emerging until light levels have fallen; this is known as light sampling.

Whiskered and Brandt's bats are difficult to distinguish in flight. Both have a rapid agile flight and fairly broad wings and often fly a regular, fixed-height 'beat' with tight turns. They both seem to have calls with a slower repetition rate than Daubenton's or Natterer's bats.

Bechstein's bats can emerge from their roosts in total darkness. They have an apparently fluttering flight but are an extremely agile species. They are woodland bats that sometimes forage in parks and gardens. They feed along enclosed vegetation mainly by gleaning, but can take prey from the ground. Their echolocation call is similar to the Natterer's bat though with a larger interval between pulses and a higher end frequency. Bechstein's bats are frequently missed using traditional bat detector survey methods because their echolocation calls are very quiet and the bats are often out of range of the detector especially when flying in the canopy. The use of acoustic lures and radio-tracking have been successful in surveys for Bechstein's bats.

Common and soprano pipistrelle bats are reasonably easy to differentiate on a bat detector from each other and also from the more dry sounding 'clicks' of the *Myotis* bats. The flight of pipistrelles appears fast and jerky as they dodge about pursuing insects, which are caught and eaten in flight. Pipistrelles also modify their call parameters slightly in response to the surrounding environment and there can be some overlap between call characteristics of Nathusius' pipistrelles in clutter and common pipistrelles in the open. The pipistrelles are the most often encountered bats especially in built-up areas.

Nathusius' pipistrelles usually emerge before the common and soprano pipistrelles, and their flight is more rapid, less manoeuvrable and with deeper wing beats. They tend to avoid built-up areas more than other pipistrelles, but otherwise use similar habitats. Their calls are also similar but can be distinguished with practice and sound analysis as they tend to use a lower frequency of maximum energy (peak frequency) and a slower repetition rate. Nathusius' pipistrelles are often found over or around fairly large standing water bodies.

Long-eared bats are sometimes called whispering bats because their calls are so quiet. If sufficiently close, their huge ears can sometimes be seen in flight. They often feed in large but enclosed spaces such as barns; they also use feeding perches and tend not to feed very far from their roosts. Their calls are louder and of lower frequency in more open areas and are very distinctive when recorded in time expansion for sound analysis. Grey long-eared bats have been reported foraging in bright white light.

Barbastelles can be fast flyers and will forage in woodland, hedgerows, meadows and riverine habitats, depending on the availability of moths, their major food source. They have a distinctive two-part echolocation call in most situations and a sonogram can aid identification of the species.

Sonogram analysis methodology

This sonogram analysis methodology was originally produced for the BCT/MTUK *Bats and Roadside Mammals Survey* by Jon Russ.

Sound analysis plays an important part in identifying bats in the field. The methodology presented here for identifying bats from their calls is based on the analysis of time-expanded recordings using computer software. The methodology does not include the identification of *Myotis* and *Plecotus* bats. The methodology is illustrated through *BatSound* but can be undertaken on other sound analysis software.

The methodology presented is a general guide to bat identification and should not be used as a precise way of determining individual species. It was originally designed for large-scale car surveys where the occasional error would not affect the overall results. It is also important to note that the habitat in which an individual bat flies has a large influence on the shape of its echolocation call and the wide range of sonogram shapes bats use is clearly illustrated below.

In order to identify which species the bat call could belong to it is necessary to determine the peak frequency of the call. To do this:

1. Highlight the call in question (see Figure 1 below) and select **Analysis-Power Spectrum** from the drop-down menu. (Note: If you cannot highlight anything try right-clicking with the mouse and selecting **Marking Cursor**.)

2. Move the **Marking Cursor** so that it hovers over the highest peak (see Figure 2) (i.e. the one closet to 0 db on the left-hand scale). Then read off the **Peak Frequency** in the bottom left-hand corner. (Note: Make sure you look at the call only. The highest peak, for example, may be at around 10 kHz which could be caused by mechanical noise. From the sonogram you should be able to see the range of frequencies covered by the calls [e.g. common pipistrelle sweeps down from about 65 kHz to 45 kHz]).

3. Identify which of the categories (A-J) the peak frequency falls into from the diagram and/or table in Figure 3.

4. Look at the example sonogram(s) in Figure 4 that relate to the categories identified at stage 3 of the process. Identify which of the shapes the call being analysed most resembles from the sonogram. For example, if you obtain a peak frequency of 40 kHz, this corresponds to both categories F and G. If looking at the sonograms in F and G the shape most resembles that of F then the call is Nathusius' pipistrelle.

There may of course be more than one bat present at the same time. This could be two individuals of the same species or two of different species.

There are other calls shapes, including those of social calls, which are not included in this key. For more information on bat identification refer to Barataud (1996), Jones and Walsh (2001), Russ (1999), Russo and Jones (2002), Schober and Grimmberger (1997) and Tupinier (1997).

Figure 1. Highlighting a call for peak frequency analysis

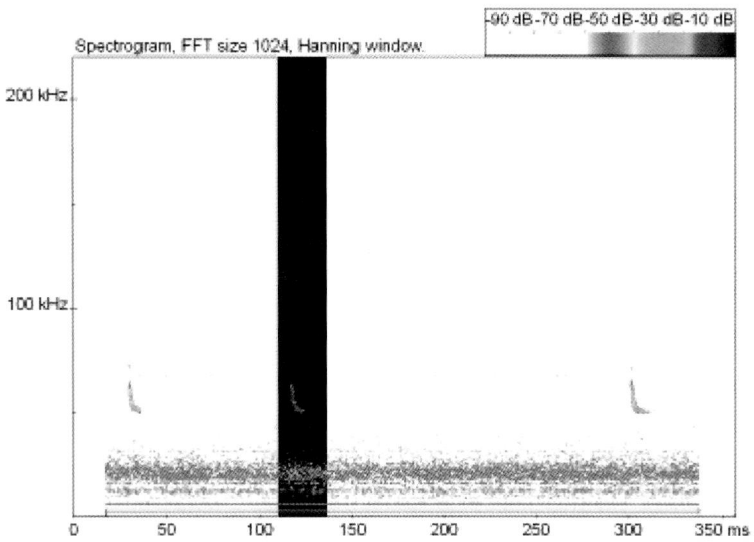

Figure 2. Using the marking cursor to read the peak frequency

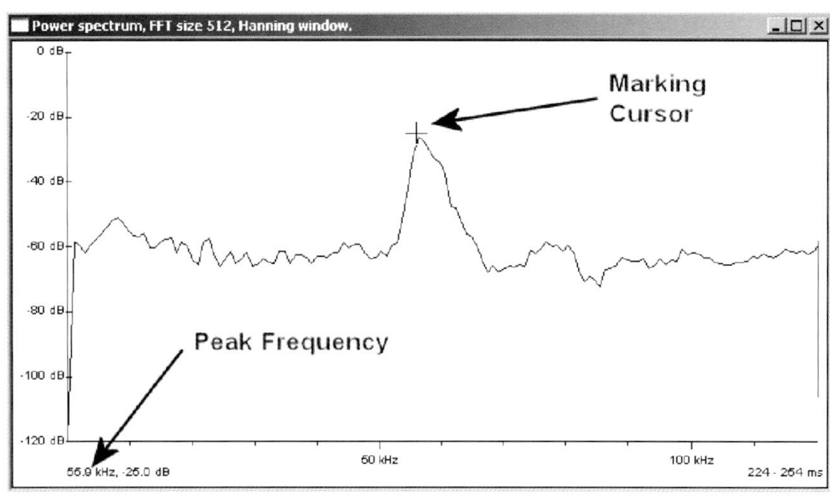

Figure 3. Diagram to illustrate the peak frequency ranges of different call categories, also given in the box at right as a list.

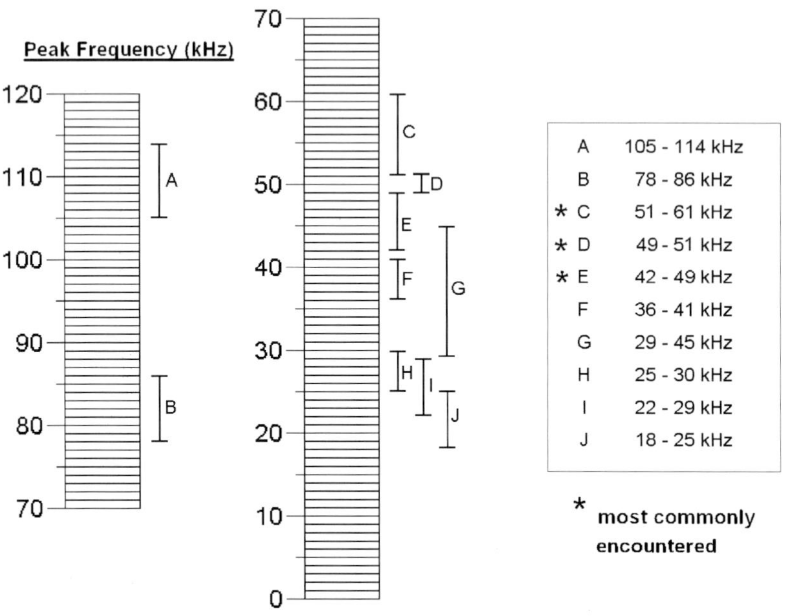

Figure 4. Example sonograms for each of the categories A-J illustrating calls of different bat species.
Note that the calls represent SINGLE calls emitted by these species isolated from a sequence and NOT a sequence of calls!

Figure 4a Examples of lesser horseshoe (A) and greater horseshoe (B) echolocation calls.

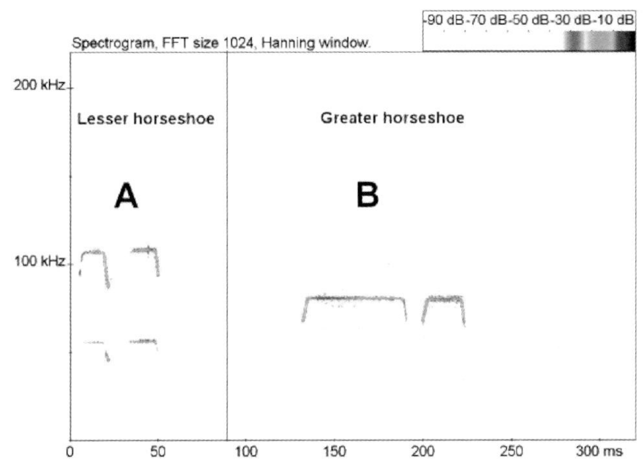

Figure 4b Examples of soprano pipistrelle (C), unidentified pipistrelle (D) and common pipistrelle (E) echolocation calls.

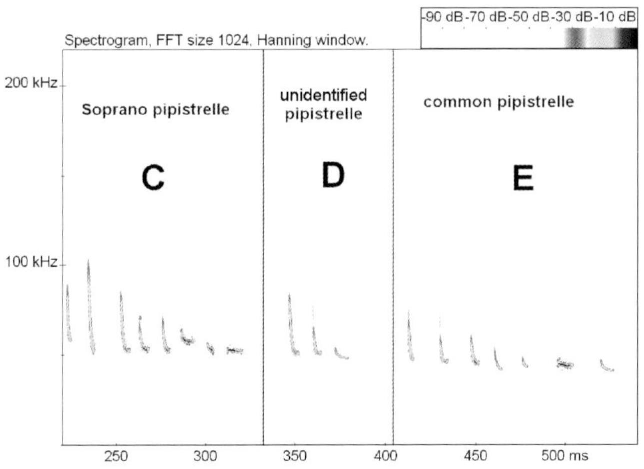

Figure 4c Examples of Nathusius' pipistrelle echolocation calls (F) and barbastelle echolocation calls (G).

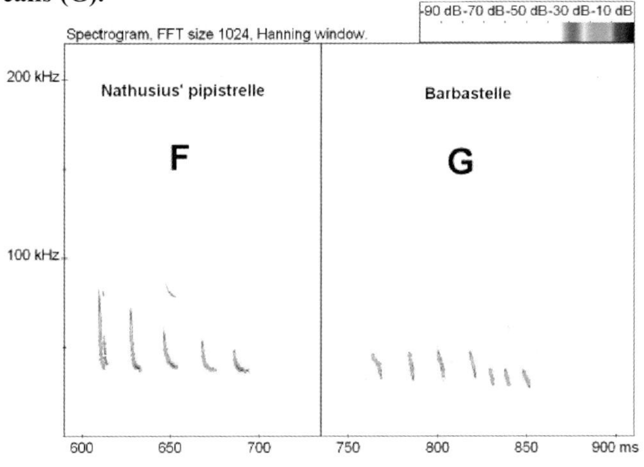

Figure 4d Examples of serotine echolocation calls. As calls become more CF and 'flattened' (to the right of the diagram) the more likely it is that the identification is correct if the peak frequency is around 26 kHz.

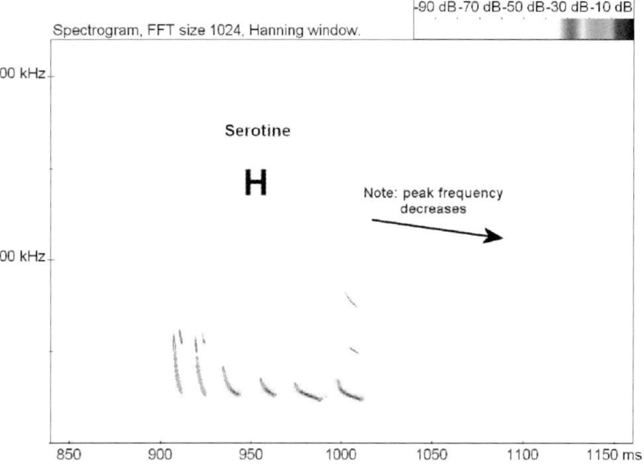

Figure 4e Examples of Leisler's bat echolocation calls. As calls become more CF and 'flattened' (to the right of the diagram) the more likely it is that the identification is correct if the peak frequency is around 23 kHz.

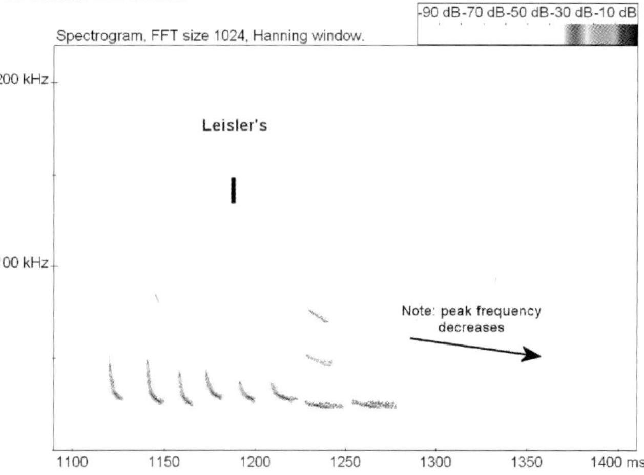

Figure 4f Examples of noctule echolocation calls. As calls become more CF and 'flattened' (to the right of the diagram) the more likely it is that the identification is correct if the peak frequency is around 19 kHz.

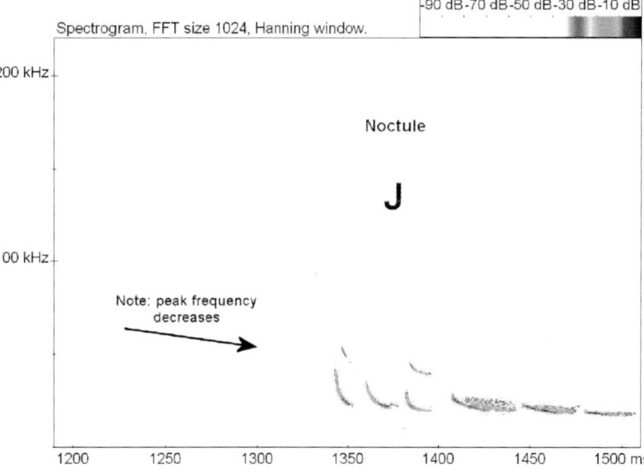

Appendix 2

Automated bat detector systems

The Anabat is an example of a complete system for the remote detection and recording of bat calls. Extensive use has been made of the Anabat system so some details on its use are provided. Other similar systems are coming on to the market, such as from Stag and Pettersson.

The Anabat hardware consists of a frequency division detector, a real-time analysis and storage module and software (*Analook*, see Section 5.5.1) for the analysis of recordings. The hardware is available as a separate detector and analysis/storage unit (Anabat II detector and ZCAIM CF storage unit) or as a single unit (SD1). A GPS unit can also be connected to the system.

The hardware performs a zero-crossing analysis of the frequency divided output from the detector and stores the resulting data on a compact flash card; the actual sound signal is not recorded. The card data are downloaded to a computer and the bat calls are displayed as frequency-time curves from which the species of bat can usually be identified.

The Anabat system is particularly suitable for remote monitoring because of the compact data storage, typically one or two megabytes (MB) per night, and the facility to run the system from an external 12V battery. The system can be programmed to switch on and off each night, and can be left unattended for many nights at a time. From the point of view of analysis, the Anabat system has a sensitivity dial, and when correctly set it will only record when bat calls are detected. Other systems record continuously, regardless of bat calls, involving the surveyor taking time to go through many hours of 'silence' during sound analysis.

Other systems can be put together using existing hardware; for example, the University of Leeds have used a system with a Batbox III, sensitive, low noise heterodyne detector and a Skye counter which counts bat passes by triggering on the first call of a pass and de-activating for pre-set time (e.g. 0.5 s) so as not to count each bat more than once. The cumulative number of calls is output to an external logger. A lead-acid rechargeable battery (12V, 4Ah = 7 days) powers the Batbox III, which is the component that limits the time the system can be left unattended. A waterproof case holds the entire system with an 11 mm diameter hole drilled into the case in front of the detector's microphone. Since it is tuned to a single narrow frequency bandwidth, it will not detect all species. For further information, see Glover and Altringham (2007).

In addition to heterodyne systems, a Batbox Duet in frequency division mode has been used to monitor greater and lesser horseshoe bats with a minidisk recorder for overnight recording (Glover and Altringham, 2007). Also, Pettersson D240x time expansion detectors have been connected to VOR walkman/Dictaphone/MP3 recorders which provide very good recording quality, although battery lifespan is short (1 night) (P. Reason and D. Wells pers. comms.)

Appendix 3

Glossary of bat terms

Autumnal swarming	Bats gathering in flight at an underground site in autumn
Dawn swarming	Bats gathering in flight outside a roost before and during sunrise
Day roost	Site where one or more bats spend the day
Feeding perch	A place where a bat hangs while detecting prey or consuming it
Hibernaculum	A winter site where the bats enter torpor during hibernation
Maternity roost	A breeding roost where mothers give birth to, and care for, their young
Night roost	A site where bats rest, groom etc. between bouts of foraging
Nursery roost	As maternity roost
Pre-lactation	The state of a female before producing milk for suckling
Post-lactation	The state of a female after producing milk for suckling
Pre-parturition	The state of a female before giving birth
Post-parturition	The state of a female after giving birth
Roost	A resting place of a bat
Satellite roost	A smaller roost than a maternity roost but nearby
Swarming	Bats gathering outside a roost at dawn or in autumn
Torpor	Slowing the metabolic rate and entering a state of deep sleep
Transitional roost	An occasional roosting site usually used in spring and autumn before and after using a maternity roost
Volant	Able to fly

Appendix 4

List of abbreviations used

ALGE	Association of Local Government Ecologists
BCT	Bat Conservation Trust
CBD	Convention on Biological Diversity
CCW	Countryside Council for Wales
CEnv	Chartered Environmentalist
CITB	Construction Industries Training Board
Defra	Department for the Environment, Food and Rural Affairs
DoENI	Department of the Environment Northern Ireland
EA	Environment Agency
EHS	Environment and Heritage Service (Northern Ireland)
EN	English Nature (now Natural England)
EPS	European Protected Species
FCS	Favourable Conservation Status
HPA	Health Protection Agency
HSE	Health and Safety Executive
IEEM	Institute of Ecology and Environmental Management
NE	Natural England (formerly English Nature)
NERC	Natural Environment and Rural Communities (Act)
NNR	National Nature Reserve
NPPG	National Planning Policy Guidance (Scotland)
PAN	Planning Advice Note (Scotland)
PPE	Personal Protective Equipment
PPS	Planning Policy Statement (England)
SAC	Special Area of Conservation
SE	Scottish Executive
SEERAD	Scottish Executive Environment and Rural Affairs Department
SEPA	Scottish Environment Protection Agency
SI	Statutory Instrument
SNCO	Statutory Nature Conservation Organisation
SNH	Scottish Natural Heritage
SR	Statutory Rule
SSSI	Site of Special Scientific Interest
TAN	Technical Advice Note (Wales)
VTA	Visual Tree Assessment
WAG	Welsh Assembly Government

Appendix 5

Useful websites

Organisations and government departments

Bat Conservation International	**www.batcon.org**
Bat Conservation Ireland	**www.batconservationireland.org**
Bat Conservation Trust	**www.bats.org.uk**
British Caving Association	**http://british-caving.org.uk/**
Countryside Council for Wales	**www.ccw.gov.uk**
Defra	**www.defra.gov.uk**
Department of Environment Northern Ireland	**www.doeni.gov.uk**
Environment Agency	**www.environment-agency.gov.uk**
EUROBATS Agreement	**www.eurobats.org**
Health and Safety Executive	**www.hse.gov.uk**
Highways Agency	**www.highways.gov.uk**
Institute for Ecology and Environmental Management	**www.ieem.net**
Joint Nature Conservation Committee	**www.jncc.gov.uk**
National Biodiversity Network	**www.searchnbn.net**
National Federation for Biological Recording	**www.nfbr.org.uk**
Natural England	**www.naturalengland.org.uk**
Office of Public Sector Information	**www.opsi.gov.uk**
Scottish Natural Heritage	**www.snh.org.uk**
Scottish Environment Protection Agency	**www.sepa.org.uk**
The Mammal Society	**www.abdn.ac.uk/mammal/**
The Wildlife Trusts	**www.wildlifetrusts.org**
UK Biodiversity Action Plan website	**www.ukbap.org.uk**

Book Suppliers

Abebooks (for out of print books)	**www.abebooks.com**
Alana Ecology	**www.alanaecology.co.uk**
Amazon	**www.amazon.co.uk**
Natural History Book Service	**www.nhbs.com**
Speleobooks	**www.speleobooks.com**